LIVE LONG AND EVOLVE

LIVE LONG AND EVOLVE

What *Star Trek* Can Teach Us about Evolution, Genetics, and Life on Other Worlds

MOHAMED A. F. NOOR

PRINCETON UNIVERSITY PRESS
Princeton and Oxford

Requests for permission to reproduce material from this work
should be sent to permissions@press.princeton.edu

Published by Princeton University Press
41 William Street, Princeton, New Jersey 08540
6 Oxford Street, Woodstock, Oxfordshire OX20 1TR

press.princeton.edu

Library of Congress Control Number: 2018944541

ISBN 978-0-691-17741-0

British Library Cataloging-in-Publication Data is available

Editorial: Alison Kalett and Lauren Bucca
Production Editorial: Kathleen Cioffi
Text Design: Jessica Massabrook
Jacket Design: Kerry Squires
Jacket Image: Olga Ovchinnkova / Alamy Stock Photo
Production: Jacqueline Poirier
Publicity: Sara Henning Stout
Copyeditor: Maia Vaswani

This book has been composed in Futura Std & Sabon Next LT Pro

Printed on acid-free paper. ∞

Printed in the United States of America

10 9 8 7 6 5 4 3 2 1

CONTENTS

PREFACE

Another biology professor asked me recently why I bothered lecturing about biology using science fiction rather than just teaching biology directly. Biology aficionados are satisfied with learning their subject directly, such as through reading popular biology books, taking online courses, watching television stations like Animal Planet, observing animals in the wild as a hobby (e.g., bird watching or bug collecting), and otherwise actively seeking to learn more about life on Earth without bothering with fiction. However, their active approach to learning biology is not one shared by everyone—many other people may become interested in the subject of biology but do not actively pursue learning more. Instead, they focus on their other interests and hobbies.

My aim with this book is to pique people's interest in biology and have them learn more about the subject by leveraging a different medium in which they may be already interested: science fiction. Science fiction captures readers' and viewers' imaginations with portraits of what may be possible, but it does so with a grounding in real scientific principles rather than relying on magic and mysticism. Magic and mysticism also capture the imagination but they fall into the somewhat overlapping category of "fantasy." Because of its grounding in science, science fiction provides a captivating springboard for an intrigued audience to learn real principles. As elaborated in the final chapter of this book, many current scientists attribute their present interest in their subject to being initially captivated by science fiction. My aim in this book is to help such a captivated audience make connections from the science fiction to the science.

While the book uses *Star Trek* as a gateway, the scenes and depictions are described here in sufficient detail that someone who has not watched the various series can still comprehend the concepts presented (both in reality and as presented in the series). One possible use for this book is for a faculty member to teach a biology course on these principles using this as the assigned text. Students could watch the episode mentioned in a section as optional "homework" for the class, but such viewing would not be necessary for discussion since the relevant scenes are recounted here. Personally, I love such courses, either as the professor or as the enrolled student.

Why *Star Trek* in particular? First, *Star Trek* offers a very large and (mostly) internally consistent volume from which to draw. Five nonanimated *Star Trek* series featuring a total of over 700 roughly 45-minute episodes have aired, and a sixth series began in 2017.* Thirteen *Star Trek* movies have also been released. Second, the *Star Trek* series often attempt to explain strange phenomena in the context of science (with varying success). While they occasionally depict magical or mystical events, most of the series is grounded in situations or observations presented as scientifically plausible. Third, precedent exists for discussing science in the context of *Star Trek*. Prof. Lawrence Krauss published national best-seller *The Physics of Star Trek* in 1995 and a revised edition in 2007. A few other related titles have also been pub-

---

* An animated series also aired in the 1970s, but I do not cover this series because some of the plots took still-greater liberties with biology.

lished, including some connected with biology. Finally, I just really love watching and discussing all of the *Star Trek* series.

Why evolutionary biology in particular? Biology is a broad field, so tackling all aspects ranging from biochemistry to neurobiology to physiology to ecology would make a book that is both long and encyclopedically tedious. Evolution is an area rife with public misunderstanding and misconceptions, particularly in the United States, so I am eager to focus on this area. Exploring evolution as it happened on Earth allows readers to understand both the extent of diversity on the planet and logical explanations for how that diversity came to be. Evolution is especially significant to *Star Trek*, where the *USS Enterprise* is tasked with "seeking out new life." Modern evolutionary studies also connect heavily with an area of great recent interest by the public: genetics. Finally, I just really love studying and teaching evolution.

Like Prof. Krauss's book on physics, this book does not catalog areas where *Star Trek* writers made errors in their depiction of evolution. Instead, my approach is to use scenes and depictions from the various *Star Trek* series or movies to introduce basic concepts and modern research in these exciting areas. My coverage of both evolutionary biology and *Star Trek* is far from comprehensive, but I am nonetheless able to cover many of the basic principles that I teach to my introductory genetics and evolution class at Duke University. The first chapter has a slightly different focus from the rest: defining and characterizing life and describing some extreme forms of life on Earth in the context of what we might see on other worlds ("exobiology") and do observe in *Star Trek*. Chapter 2 describes the evidence for evolution on Earth and evaluates three explanations for extraterrestrial humanoid life presented in *Star Trek*. In chapter 3, I present an introduction to genetics on Earth and depictions of genetics and genetic engineering in the series. Chapter 4 uses the genetics concepts from the previous chapter to explain basic evolutionary processes like natural selection and genetic drift. Chapter 5 delves into various forms of reproduction both within species and between species on Earth and in *Star Trek*. This chapter gets at the question of the surprising number of species "hybrids" (e.g., Mr. Spock in the original *Star Trek* series has a Vulcan father and human mother) depicted in the various *Star Trek* series. The final chapter discusses the importance of basic science and its depictions in science fiction, and how such depictions have inspired interest in science. To broaden the coverage, I also discuss a few extra "just-for-fun" related topics in the appendix of the book.

I hope readers enjoy the book and are inspired to look up some of the references cited or other resources to learn more about evolutionary biology. I would not mind if a few decided to go (re)watch some *Star Trek* episodes or movies, too. Live long and prosper!

ACKNOWLEDGMENTS

I thank Alison Kalett at Princeton University Press for entertaining this idea, for many brainstorming sessions about style and content, and for helpful feedback on the content. This book would not have come to fruition without her enthusiastic energy. Several friends and colleagues provided helpful feedback on draft chapters, including Megan Elise, Chris Friedline, Norman Johnson, David MacAlpine, Jennifer Forrest Meekins, Sandy Moore-Furneaux, Rachel Newcomb, Maria Orive, Christopher Roy, Kieran Samuk, Amy Schmid, Eric Spana, and John Troan. Stephen Craig also provided additional feedback on some concepts. I thank my employer, Duke University, for allowing me to indulge in this activity, and also Michael Nachman, the Nachman laboratory researchers, and the Miller Institute for Basic Research in Science at the University of California at Berkeley for hosting and supporting me as I completed this project. Most of all, I thank my family: my wife Julie, daughter Megan, and son Adam. They put up with endless dinner-table discussions about interesting tidbits I discovered while researching the science in the book or while rewatching hundreds of hours of *Star Trek*.

ABBREVIATIONS FOR
STAR TREK SERIES

TOS	*Star Trek* (The Original Series)	Aired 1966–69
TNG	*Star Trek: The Next Generation*	Aired 1987–94
DS9	*Star Trek: Deep Space Nine*	Aired 1993–99
VOY	*Star Trek: Voyager*	Aired 1995–2001
ENT	*Star Trek: Enterprise*	Aired 2001–5
DIS	*Star Trek: Discovery*	Aired 2017–

CHAPTER 1

"TO SEEK OUT
NEW LIFE . . ."

The opening sequences of both *TOS* and *TNG* mention that their mission is "to seek out new life." While many viewers think of this mission in the context of other humanoid life forms depicted in the series, such as the Vulcans or Klingons or Andorians, fewer think of it in the context of entirely unfamiliar forms. Is such unfamiliar life likely, and what might a "new life form" look like? The first section of this chapter examines briefly the question of what "life" actually means and begins to examine its probability of occurrence. Many biology textbooks provide lists of characteristics of living organisms, but exceptions to some items in these lists abound even on our own planet. Might we expect similar forms to have arisen elsewhere in the universe? Furthermore, these lists are artificial in the sense that they were made based on observations of known organisms on Earth rather than derived from fundamental principles of biology and chemistry. Much of life is, for example, water and carbon based, but are these properties general to life or idiosyncratic to the observed single set of related life forms on Earth? The second section of the chapter goes into several properties associated with life on Earth, and considers what alternatives may be possible.

TNG, Season 6, Episode 9, "Quality of Life"

Lieutenant Commander Data is an android but has been considered throughout the series to be alive and self-aware. In this episode, he begins to question whether other machines he encountered are alive. He goes to the ship's chief medical officer, Beverly Crusher, and asks what the definition of life is. She answers that "the broadest scientific definition might be that life is what enables plants and animals to consume food, derive energy from it, grow, adapt themselves to their surroundings, and reproduce." Data is dissatisfied with this answer, noting fire "consumes fuel to produce energy, it grows, it creates offspring." He then elaborates on an exception in the other direction, noting about himself that he does not grow or reproduce yet is considered alive. He further raises the question of whether something specific transpired to endow him with life between when he was merely component parts and when he became alive.

DEFINING LIFE

Much science-fiction writing and film (*Star Trek* or otherwise) considers the possibility of life on other worlds. We all ponder how likely extraterrestrial life is to arise and what form such life might take. To answer these questions, we need some idea of how to define life so that we may determine its likelihood and recognize it when we come upon it. To define life, we humans naturally first consider forms of life with which we are most familiar: life on Earth. We know that animals, plants, and fungi are alive. We are similarly convinced that various microscopic forms including bacteria, amoebae, and the parasite that causes giardia are alive. What attributes do these forms share? The textbook list of traits associated with life is similar to the ones that Doctor Crusher laid out in *TNG* "Quality of Life" above: acquiring or producing energy, some level of internal organization, maintenance of a constant internal environment, growth, reproduction, response to stimuli, and ability to adapt over generations.[1] Physicist Erwin Schrödinger highlighted in 1944 that living matter evades the decay to equilibrium.[2] Nonliving matter (whether never alive or now dead) tends to move to equilibrium with its environment: reaching a similar temperature, not growing or moving unless impacted by outside forces, etc. Life forms expend energy to remain distinct and out of equilibrium relative to their environment.

Let me use a bacterium that causes pneumonia, *Streptococcus pneumoniae*, as an illustrative example. It appears in the form of individual cells as units of life. Each cell takes up simple carbohydrates from its environment to use for energy; each has a defined structure with a wall and several specific surface proteins to maintain the internal environment; each has internal structures including a chromosome containing instructions for maintenance and reproduction; each grows and reproduces; each produces specific proteins in response to antibiotics; and several cell lineages have evolved over time, in some cases, unfortunately for us, to become resistant to commonly used antibiotics. Unambiguously, *Streptococcus pneumoniae* cells are alive.

While seemingly simple in some cases, defining life becomes a challenge when there is a mismatch between the list of characteristics and our instincts on whether something is alive. As Data noted, fire converts matter in its surroundings to energy and grows. A small spark from a fire can allow a separate fire to emerge, analogous to reproduction. Nonetheless, no biologist argues that fire is alive. Our homes have internal organization and a constant internal environment in some respects, but we do not perceive them to be alive. On the other side, many organisms we presume to be alive—in addition to the android Data—are also unable to reproduce, such as the mule (the sterile hybrid offspring of a horse and donkey: see chapter 5). Besides reproduction, other attributes associated with life are not "essential" for life: we would not declare an individual "nonliving" solely because it failed to grow or if it reproduced as completely identical clones so that there was no potential to adapt over time.

There are also gray areas. Are viruses alive? Viruses require nonvirus host cells for reproduction, so they cannot reproduce independently. They do not generate or store energy but instead rely on host cells for energy for all functions. Some scientists have argued that these properties make them too dependent on other life forms to be considered alive. However, many living organisms are dependent on other individuals or species for their lives or reproduction. Medical professionals often talk about "killing viruses," which implies that they are alive. Recent work shows individual viruses even exhibit a form of chemical communication that affects the behavior of other viruses.[3] Overall, biologists are split on the question of whether viruses are alive or merely natural replicators that capitalize on and influence other living organisms.[4] Viruses are not the only gray area. Transposable elements (DNA sequences that insert themselves into genomes and then make copies of themselves) and prions (misfolded proteins that change other proteins near them to their misfolded state) also self-replicate in a sense, but they are considered to be further from "living" than viruses.* As these ex-

---

* Prions are associated with various human and other mammalian diseases, and reference was made in *DS9* episode "Business as Usual" to leveraging prions as a weapon to kill 20 million people.

amples illustrate, there is no simple solution to defining life. In essence, "life" is an imprecise term used to define having many of a suite of particular traits, but with no specific number or absolute requirements of which traits from the suite must be included.

Despite this uncertainty, some very reputable biologists, including Nobel laureates, have argued that defining life precisely is not necessary for us to study the likelihood of life to arise or its possible origins.[5] We still know many important components, and those can be researched individually or in groups. Life as we know it arguably has a chemical basis; venturing outside that constraint generally falls into the realms of philosophy and religion. With that in mind, some scientists have suggested that life can be described as a sustained chemical system that undergoes self-reproduction with the potential for some change over time (i.e., evolution).[6]

GENERATING NEW LIFE

If the simplest form of new life is a sustained chemical system like that described above, such life has already been constructed in the laboratory using building blocks from existing life: specifically synthesized fragments of the nucleic acid RNA (also discussed in chapter 3). The first life on Earth may well have used RNA for heredity, since, unlike DNA or proteins, RNA has the ability to both transmit hereditary information and carry out some vital functions of the cell. Indeed, RNA is involved in the transmission of genetic information in some present-day viruses, such as HIV. Self-replicating combinations of RNA "instructions" have been assembled in the laboratory that, in the presence of the appropriate raw materials, make more copies of

themselves without any added directions or machinery (e.g., enzymes).[7] More recent studies have given these RNAs the ability
to produce other types of functional molecules.[8] Hence, some
of the most plausible models for early life on Earth include an
RNA-based phase, making this particular example especially
interesting for understanding the history of life on our own
planet.[9] Still, self-replication of genetic material alone does not
make a "cell" as we know it today—metabolic processes must
also occur, and the cell-replication machinery must remain
physically distinct from its immediate surroundings. On the
latter point, recent progress has been made in describing how
membranes also may have evolved, to keep cells distinct.[10]

Would the raw materials for life have been available on early
prelife Earth, though? A famous experiment, published in 1953
by Stanley Miller, showed that amino acids, the building blocks
of proteins, can form naturally without preexisting life in the
presence of hydrogen, ammonia, methane, and water, when
exposed to sparks (analogous to lightning).[11] While we do not
know with certainty the exact chemicals or their concentrations
on Earth before life formed, these compounds are widespread in
the universe, and related results were found in later studies using different chemicals (e.g., hydrogen sulfide) or different conditions. Additionally, numerical models suggest that long RNA
molecules, potentially able to initiate primitive life, could have
formed more than four billion years ago on Earth in the conditions present at the time.[12] Altogether, the potential seems high
for having the known raw materials for Earthlike life arise spontaneously in the universe.[13]

Much of the text above focuses on the question of how life
on Earth might have arisen—understanding a specific instance.
If one is considering life on other worlds, we must focus on the
more general question: how might life arise? While the studies

described above provide elegant proof of principle of the origin of basic life components from nonlife on an early Earth, they may not reflect the potential for life on other worlds accurately.

As such, a few scientists have taken an even more basic physical view in exploring the potential for the conditions associated with life to arise. Earth is an "open system," in that not only do interactions happen between organisms and resources on the planet, but also energy is continually being provided to the planet via radiation from the sun. Many systems tend to spread energy out over time (increasing entropy, as per the second law of thermodynamics), but open systems can divide energy unequally since they are influenced by and can influence their surroundings. Under such conditions, and if surrounded by a liquid or gas (e.g., our planet's oceans or atmosphere), theory suggests that matter may often gradually restructure itself so as to dissipate greater amounts of heat energy. Some physical scientists have argued that self-replication (reproduction) may achieve this outcome, since replication dissipates energy in an irreversible manner (i.e., one organism is more likely to replicate into two than two organisms are to fuse into one).[14] General thermodynamic definitions of evolution predate this particular model, and certainly the applicability of this argument to the origin of life in particular is tenuous. However, the argument described above adds new dimensions, suggesting why processes associated with life may be somewhat likely to arise from basic principles of physics.

Taking all of the above together, "life" on other worlds may be simple and reasonably probable to exist but could arguably be chemical processes quite unlike the humanoid aliens observed in much of *Star Trek* or other science fiction. Such forms may be extremely difficult to notice, even from very close, and simple "scans" from space, such as those conducted in *Star Trek*, may easily miss them. Instead, *Star Trek* devotes much attention to the "sentience" of life across the series: the crews often struggle to determine whether organisms that they encounter are self-aware. Undoubtedly, self-awareness would indicate that an entity is alive, but the vast majority of organisms we know or predict would not have this characteristic.

NEW LIFE IN *STAR TREK*

As in the example at the start of this section, *Star Trek* does consider the potential for life among constructed forms, in which the life-related processes may be electrical rather than chemical and some exceptions to the "characteristics of life" may apply. Two examples used in the series regularly are the *TNG* android character Data and the *VOY* holographic doctor. Both characters are essentially animated computer programs, yet both have most or all of the characteristics of life and are even self-aware. In fact, both reproduced in some sense: in *TNG* "The Offspring," Data built a child (Lal) using neural transfers from himself, while in *VOY* "Real Life," the holographic doctor created a holographic wife and children. Are these truly "offspring"? Counselor Troi emphasized to Captain Picard, "Why should biology rather than technology determine whether it is a child? Data has created an offspring. A new life out of his own being."[†] Can a "manufactured" form be considered alive? Multiple religions suggest that existing species (including humans) were designed by a living creator, and no one argues that such a premise would mean the products are "not living." Hence, constructed forms can be considered alive in principle, and we may be close to producing

†   *TNG* "The Offspring."

human-made forms that can be considered alive. One might wonder if perhaps the first new life we encounter may be a form that either we made ourselves or that others manufactured and sent out into the cosmos.

Nonetheless, when biologists discuss the origin of life, they typically consider life arising from raw materials of nonlife, and not initiated or produced by extant living forms. In that regard, while androids or computer programs may fit the definition of life, they did not arise from nonlife. I suggest a separation of "origin of life from nonlife" and "origin of new life from existing, albeit different, life." Although many *Star Trek* episodes are devoted to the latter, fewer focus on the former. One passing reference was made in the *DS9* episode "Playing God" to potential new life evidenced by "nonrandom thermodynamics," akin to the basic physical view discussed above. However, few (if any) other references exist across the series.

HOW MUCH SHOULD LIFE ON EARTH RESEMBLE EXTRATERRESTRIAL LIFE?

Life forms on Earth are made from carbon-containing compounds and use water for their biochemical reactions. Many Earth life forms thrive at temperatures between 5°C and 40°C and release energy from respired oxygen. Should we expect extraterrestrial life to share such characteristics? The problem with extrapolation is that all of life on Earth is related in a strict sense: we share a common ancestor with every life form known on this planet (a topic discussed at greater length in chapter 2). Hence, while we know many life forms on Earth are water and carbon based, all of those forms are nonindependent, creating a problem for predicting what we may see elsewhere. This problem is analogous to predicting the characteristics of a "sport" having only known soccer and some sports derived from it (e.g., American football). Knowing these sports, one might then predict "sports" to involve teams, scores, and putting an object into a goal defended by an opponent. While these attributes apply for some sports, like basketball or hockey, how well would such predictions apply to fencing or surfing? What about karate or boxing? Many sports have some means of scoring (albeit measured in very different ways) but do not involve putting objects into defended goals.

Through the next sections, we will consider a few specific characteristics of life on Earth, variations thereof observed on Earth, and, when possible, arguments for whether they may be typical or exceptional on other worlds. We will also look at whether and how variations were explored in the *Star Trek* series. In principle, one should predict that life forms on other worlds and in *Star Trek* should be *more* diverse than life on Earth if they do not share evolutionary ancestors with each other and with Earth's life forms (but see chapter 2)—analogous to how unrelated people often exhibit greater variation in features than do genetically related people within a single family. Some of this predicted greater diversity is indeed reflected in the series.

WATER

TNG, Season 1, Episode 18, "Home Soil"
Some crew members find a nominally "inorganic" (not containing known building blocks of life such as carbon compounds) life form. It has no carbon, but it has silicon, germanium, and other elements. The life form eventually communicates with the crew and refers to the humans on the ship as "ugly, ugly, giant bags of mostly water."

The alien's point is fair: humans are mostly water, as is most life on Earth. Much of the search for extraterrestrial life also focuses on water. Researchers with space telescopes emphasize investigations of planets in what they define as the "habitable zone" or "Goldilocks zone": planets "not too close" yet "not too far" from their star for water to exist in liquid form on the surface of a rocky planet. In 2009, NASA launched the *Kepler* space telescope to seek and estimate the abundance of approximately Earth-sized planets in or near the habitable zones of their stars. Kepler-22b, Kepler-62f, and Kepler-452b are examples of such planets inferred.[15] Many others have been found by other means: for example, seven planets with masses similar to Earth appear to orbit the dwarf star TRAPPIST-1 in its habitable zone,[16] and planet LHS 1140b is in the habitable zone of its star and appears rocky and roughly Earth sized.[17]

Why liquid water? Water is the primary solvent for all life on Earth. Let me start with some very basic chemistry. Solvents are often in liquid form, and other chemicals dissolve in them. Solvents can get substances used in chemical reactions into the same phase so that molecules can collide with each other and react. Such reactions facilitate the sustained chemical systems we consider life, as discussed in the first section. While solvent-free reactions exist, having a solvent often greatly simplifies having sustained chemical reactions.

There are reasons that water may be generally, rather than rarely, the solvent used for life. The simplest reason has to do with sheer abundance. Water is composed of two hydrogen atoms and one oxygen atom. Hydrogen is, by far, the most abundant element in the universe. Oxygen is the third most abundant element in the universe. The second most abundant element, helium, is the least reactive in the periodic table, and therefore would not likely be a component of a solvent. Hence, by sheer abundance and simplicity (three atoms), the constituents of water easily beat out all competing solvents.

Water has several other favorable properties. It is liquid at a broad range of temperatures. Water is a "polar" molecule, in that the oxygen atoms have a partially negative charge and the hydrogens have a partially positive charge, due to differences in electronegativity between those two elements. This property makes water particularly good at dissolving salts and other molecules that are polar by nature; in fact, water is capable of dissolving more substances than virtually any other liquid. Molecules that do not exhibit polarity, such as fats and some other molecules rich in hydrocarbons, do not dissolve in water. Nonpolarity is also sometimes beneficial, since nonpolar molecules in water are sometimes forced to interact with each other rather than with the solvent. These features work very well for living systems on Earth and are advantageous for organizing cell walls, mediating interactions among or folding of proteins, etc.[18]

On the other hand, life on Earth has evolved around the use of water as a solvent for roughly four billion years, so water may now appear to be more "ideal" than it was when life first arose. In other words, while life on Earth today works splendidly with water but much less so with, for example, liquid ammonia, this contrast may result simply because of four billion years of natural selection optimizing living forms for use with a less-than-ideal water solvent rather than water having been ideal at the start.

In this regard, there are reasons that perhaps water may not be ideal as a solvent for life on other worlds. First, its liquid form is associated with rather high temperatures (0°C–100°C at Earth's atmospheric pressure). The vast majority of the universe and, presumably, many planets are far colder than that. As implied above, atmospheric pressure is also important: liquid water cannot persist on Mars, for example, because ice transitions directly to water vapor given the low atmospheric pressure. Water also damages or degrades some important molecules associated with life, such as some proteins, DNA, and RNA, but terrestrial life has adapted to mitigate this damage through repair or through producing more molecules than needed so damage to a subset can be tolerated.

Alternative potential solvents exist, including ones that are liquid at much colder temperatures. One of the most likely alternatives may be liquid ammonia. Like water, it is polar and can also dissolve many known organic compounds. It maintains a liquid state at −78°C to −33°C, but can be liquid at much broader temperatures if at high pressure. For example, ammonia is liquid from −77°C to 98°C at 60 times the atmospheric pressure of Earth, which is less than the pressure on the surface of Venus, for instance.[19] Ammonia is abundant across the universe—the atmospheres of many of the outer planets in our solar system have it. Saturn's largest moon, Titan, may even have subsurface ammonia-enriched liquid water oceans,[20] which could function as a solvent mixture for life-related processes. Other solvents may work for life at distinct temperatures or pressures, such as high-pressure carbon dioxide[21] or methane,[22] or life may even possibly—though arguably more problematically—involve reactions in gaseous or solid phases.

We predict that alien life forms should sometimes also use water as a solvent, but they might also sometimes utilize alternative solvents. How well does life in *Star Trek* reflect this prediction? For simplicity in terms of scientific consideration throughout this entire chapter, I set aside any noncorporeal life forms from discussion, even though several have appeared in all of the *Star Trek* series.[‡] I similarly exclude species that readily change form or

---

‡ Some examples are the *TOS* Medusans, *TNG* Calamarain, *DS9* Prophets, *VOY* Paxans, *ENT* wisps, and *DIS* Pahvans.

between states (e.g., solid to liquid) or between matter and energy.§ Finally, I only consider organisms that exist in normal space, not in fictional alternatives like "subspace" or "fluidic space."

· Alternatives to water-based life are not addressed directly in *Star Trek*, and this absence may reflect a real-world bias for seeking extraterrestrial water and seeking extraterrestrial life. In defense of the writers, the series tend to present characters spending most of their time in atmospheres, temperatures, and pressures not very different from what we experience indoors. As noted, many of the alternative solvents would be more likely to work in very different conditions, much higher pressure or much lower temperature for example, and therefore we do not expect to encounter them (or organisms based on them) in the series given the frequent depictions of human-comfortable conditions. Further, just as many nonwater solvents (e.g., alcohols, phenols) are toxic to life forms that we know (e.g., many, albeit not all, microbes), water may be disruptive or even toxic to non-water-based life forms. Hence, if the characters in the series are often present on planets with Earth-like atmospheres, they are intrinsically unlikely to encounter the full range of possible alternative life forms. I appreciate that NASA prioritizes Earth-like conditions in seeking new life since we do not yet know if *any* extraterrestrial life exists. However, because extraterrestrial life has been found in *Star Trek*, the *USS Enterprise* may be well advised to visit more planets with non-Earth-like conditions to better explore the boundaries of "new life," as its mission purports to seek.

As early as 1962, biochemist and famous author Isaac Asimov wrote an elegant essay expressing his frustration with science fiction not being imaginative enough about life (e.g., always por-

§ Recurrent examples being *TNG* Q and *DS9* Changelings.

trayed as water/carbon based) and including a table showing hypothetical "life chemistry" combinations going from extremely
high temperatures to near absolute zero.[23] His table lists carbon-
based proteins and nucleic acids—for example, DNA and RNA—
in water interacting well in typical Earth temperatures, fluoro-
carbons in sulfur working at higher temperatures, and lipids in
methane working at lower ones. Chemicals and their solvents
must be considered as combinations: the solvent and the elemental basis of solid life need to interact appropriately and be
able to do so in the specific environmental conditions of, for example, temperature and pressure. While acknowledging this fact,
let me consider alternatives to carbon and other characteristics
of life on Earth as well.

CARBON

TOS, Season 1, Episode 25, "The Devil in the Dark"

On a mining colony, the crew encounter a creature (identified later as a "Horta") that appears to be made of rock.
To attempt to make the humans leave the mining colony,
the Horta steals the circulating pump that provides air (including oxygen) to the miners. The *Enterprise* crew begin
to speculate about this mysterious life form, particularly
after learning of silicon nodules found in the mines. Science officer Spock notes, "Life as we know it is universally
based on some combination of carbon compounds, but
what if life exists based on another element? For instance,
silicon." Skeptical, Doctor McCoy counters, "Silicon-based
life is physiologically impossible, especially in an oxygen
atmosphere." After adjusting their scanning devices (tri-
corders) to look for silicon, they are able to find the Horta,

and when they injure it with their weapons, they find the piece cut from it is indeed made of silicon-based fibers.

Carbon is intrinsically related to life on Earth, so much so that it essentially defines life as we know it. The term "organic" is defined simply as relating to or derived from living organisms, but "organic compounds" and "organic chemistry" refer to matter that contains carbon atoms. Fundamental molecules of life on Earth are all carbon based: fats, carbohydrates, proteins, DNA, RNA— all of these have carbon atoms forming a so-called "carbon back- bone." Carbon is a relatively small, lightweight atom. Because of its position in a central column of the periodic table, carbon is able to bond with up to four other atoms and is quite versatile in being able to produce diverse structures (long, branching, or ring- like, for example). In much of the biochemistry on Earth, carbon atoms are bound to other carbon atoms as well as to hydrogen and oxygen, and sometimes nitrogen, phosphorus, and sulfur. For many reactions associated with metabolism or other life-related processes, carbon and/or hydrogen will dissociate from one of the other elements, and new bonds are formed soon thereafter.

Carbon is also the fourth most abundant element in the uni- verse, making it a good candidate for life. As noted earlier from Stanley Miller's classic experiments and various follow-up stud- ies with varying conditions, nonbiological formation of many carbon-based compounds, such as amino acids, has occurred under laboratory conditions,[24] again suggesting the feasibility of a similar process having occurred on other worlds. Indeed, scientists have even isolated extraterrestrially formed carbon- based amino acids,[25] as well as components of DNA and RNA,**

** *VOY* "Body and Soul" mentioned "primitive strands of DNA" taken from a comet.

from meteorites,[26] further demonstrating that these forms have arisen on other worlds and may be associated with life elsewhere. Finally, many carbon molecules used in life exhibit a geometric property called "chirality" (meaning they occur in two geometric forms that are mirror images of each other, like a person's hands), and most naturally occurring sugars differ in their chirality from naturally occurring amino acids. This chirality allows for particular kinds of useful molecular interactions that are used in life.

Given all of these good properties, are there alternatives? As described in the example above, *Star Trek* explored one of the most often considered alternative elements: silicon. Situated immediately below carbon on the periodic table, it also shares the ability to bind with up to four other atoms and is almost as small. While carbon is more abundant in the universe at large, silicon is more abundant than carbon in the Earth's crust. However, while carbon makes long chains with itself or binds with other elements, giving it a lot of flexibility, silicon binds more tightly to other elements like hydrogen or oxygen than to itself. Long-chain silicon–silicon molecules may decompose in water because silicon preferentially binds to the other atoms. Silicon molecules are also less likely to exhibit the chirality that carbon molecules do, diminishing flexibility. Most organic molecules do not contain silicon despite its abundance in Earth's crust, although some laboratory evolution studies have produced bacteria that form molecules with carbon–silicon bonds across a range of conditions,[27] illustrating some potential for "use" of silicon in life even if it were not the primary life element. (Interestingly, a silicon-based parasite was manufactured in *VOY* "The Disease," which aired long before the real-world study described above.)

The tight bonding of silicon to oxygen would make silicon as a primary component of life very challenging on modern Earth.

Presumably it was for this reason that Doctor McCoy noted silicon-based life is physiologically impossible in an oxygen atmosphere. This bonding is what makes silicon dioxide a rather unreactive solid—in forms like quartz or sand—on Earth, whereas carbon dioxide is a gas that reacts more readily. However, Doctor McCoy did not address one related possibility: life involving repeating units of silicon and oxygen ("silicones"), perhaps with carbon too, may work well in worlds with much higher temperatures than are typical on Earth—potentially, in some respects, even better than our carbon-only-based forms.

Overall, it does seem that carbon has properties that make it unusually good for facilitating life, at least on planets with Earth-like conditions and perhaps even somewhat more generally. It may be the most abundant backbone upon which life would form. Nonetheless, we cannot exclude the possibility for non-carbon-based life.

As discussed, several non-carbon-based life forms have been highlighted in *Star Trek*, with the Horta being the best-known example. When Doctor McCoy noted that such an organism could not persist in an oxygen-based atmosphere, Mr. Spock suggested that perhaps it "can exist for brief periods in such an atmosphere before returning to its own environment." Perhaps the Horta stole the pump that provided air to the human colony in this episode to reduce this oxygen exposure. Another example was the silicon-based form from *TNG* "Home Soil" mentioned in the preceding section. However, this life form seemed devoid of carbon, and yet it was hypothesized later in the episode that it used saline water in some way (either as nourishment or as a "connection" among cells). Without knowing the exact form of the silicon-based molecules, it is hard to assess how likely such an organism might be in such an environment. A silicon-based virus was found in one episode of *ENT* ("Observer Effect"), and a gigantic, space-faring, silicon-based, crystalline entity was also depicted in a few episodes of *TNG* ("Datalore," "Silicon Avatar"). However, one of the more interesting possibilities was depicted in the Tholian species (*TOS* "The Tholian Web," *ENT* "In a Mirror, Darkly"). They have an exterior made of an unspecified mineral that could be silicon based, and they live at extremely high temperatures (>200°C). Again, details are not specified, but the show writers laudably thought "outside the box" in this choice, arguably with some consideration of the underlying science.[††]

In addition to demonstrating noncarbon compounds associated with life in *Star Trek*, the episodes described above also emphasize the variation in temperature experienced by life forms (from below –250°C in outer space to above +200°C, as the Tholians experience). The next section examines variation in temperature both on Earth and in space.

TEMPERATURE

VOY, Season 4, Episode 24, "Demon"

The *USS Voyager* has come to a "demon-class planet": with a toxic, radiation-filled atmosphere and a surface temperature in excess of 220°C. Nonetheless, upon exploring the surface, they find a life form called the "silver blood."

The range of temperatures among planets even just in our solar system is enormous, from roughly –210°C on Uranus to 460°C on the surface of Venus. This raises the question of what temperature

[††]   One exception to this is Neelix's reference to a "xenon-based life form" in *VOY* "Hope and Fear." Not only is xenon a gas, but it does not bind to other elements easily. This suggestion is implausible based on what we know.

range can possibly support any form of life, known or hypothetical. As discussed earlier in the chapter, water being the solvent of life on Earth necessarily limits temperature conditions for life on Earth. I touch on the extremes of the range of temperatures experienced by living organisms on Earth only briefly because many books have been written about such "extremophiles." One important distinction to retain in this discussion, however, is the difference between "tolerance"—meaning an organism can minimally survive under such conditions without dying, but perhaps only temporarily and/or without reproducing—and the conditions under which an organism can thrive and reproduce. Additionally, the conditions on our planet today may be quite unlike the conditions when life first arose, so the temperature range under which Earth life exists today may also differ from eons ago.[28]

Some heat-tolerant "hyperthermophile" species are known to survive and reproduce at temperatures near or above 100°C. For example, the archaeon microbe *Pyrococcus furiosus*, found in deep-sea hydrothermal vents, has an optimal growth temperature of 100°C and can reproduce every 37 minutes.[29] The genus name essentially means "fireball." Another archaeon microbe from hydrothermal vents, inelegantly described as "strain 121," can grow at temperatures between 85°C and 121°C, reproducing after 24 hours at 121°C.[30] These microbes could survive periods at 130°C and subsequently reproduce when the temperature cooled slightly.

At the other end, several cold-tolerant "psychrophile" species are known that can reproduce at temperatures near freezing and survive at much lower temperatures. For instance, the marine bacterium *Moritella profunda* grows best at 2°C or lower.[31] Generally speaking though, free-living, single-celled microbes on Earth are presumed to be unlikely to be able to grow and reproduce below –20°C due to biophysical limitations associated with liquid water,[32] although water can remain liquid at lower temperatures if particular chemicals are dissolved in it. Multi-cellular organisms can potentially reproduce at lower temperatures; for example, emperor penguins can breed at temperatures of –40°C. Many organisms can survive (without growing or reproducing) at still lower temperatures. Tardigrades (also called water bears; more on these creatures in the Appendix) can survive submersion into liquid nitrogen (–196°C),[33] and a handful of individuals studied even survived a few days' exposure to the vacuum and radiation in low Earth orbit.[34] However, some far more extreme stories concerning these interesting organisms (e.g., that they can survive freezing for 100 years) may have been exaggerated.[35] Nonetheless, some bacteria were extracted from ancient ice cores dating back 120,000 years, and were able to reproduce after thawing.[36] Clearly, some Earth organisms can tolerate very long and/or very cold freezing. These observations prompt the intriguing possibility that life similar to Earth's may lie in wait on other worlds, perhaps even in our solar system.

What is observed in *Star Trek*? Many of the real-world examples described here involve microbes, and *Star Trek* rarely elaborates details on the ecology of microbes. Chief Engineer La Forge intriguingly suggested that thousands of single-celled life forms were present on "any planet's surface" (*TNG* "Time's Arrow, Part 1"), certainly implying diverse tolerances given the range of conditions across planets. Another noteworthy reference was Doctor Phlox's (*ENT* "Breaking the Ice") response to a question about whether germs can live in space. He noted there were millions of cataloged space-dwelling microbes that drift in dormant state through space and can then infect humanoids. While we know of no such species today, the existence of such microbes may be plausible given the diversity of life observed on Earth.

Several other much larger space-dwelling species were also observed in the various *Star Trek* series (e.g., *TNG* "Galaxy's

Child," *VOY* "Elogium," *DIS* "Magic to Make the Sanest Man Go
Mad"). At the warmer end, the previously mentioned Tholians
(*ENT* "In a Mirror, Darkly") require temperatures around 200°C,
and the mimetic silver blood species (*VOY* "Demon") was found
on a planet with temperatures exceeding 200°C. Interestingly,
many of the above-mentioned species are likely not water based
(the gormagander being a likely exception, *DIS* "Will You Take
My Hand"), potentially granting them the ability to thrive in
conditions outside of those observed in life on Earth. However, I
find it notable that the crews very rarely encounter alien species
requiring cooling or heating space suits that maintain the occu-
pant even 20°C above or below room temperature.[‡‡] Generally
speaking, the series tend to focus on aliens in room-temperature
environments, and the possibility for extremophiles (relative to
Earth environments) seems underexplored.

Finally, irrespective of chemical composition or temperature
conditions, the processes associated with life necessarily require
energy. The final section explores energy production on Earth
and in *Star Trek*.

ENERGY

TNG, Season 4, Episode 16, "Galaxy's Child"
The *USS Enterprise-D* encounters a space-faring life form
orbiting a planet. It sends an energy field over the ship,
and the crew are forced to shoot it to escape. However,

[‡‡]   The Nyrians in *VOY* "Displaced" are one exception since they preferred a
temperature of 45°C (~20°C above room temperature). The Breen in *DS9* are another possible exception since they wear "refrigeration suits," but Weyoun in "The
Changing Face of Evil" suggested the suits may not be related to temperature.

the life form had an offspring within its body, and the offspring emerges after the mother dies. The baby alien follows the ship, comes into contact with the hull, and as Science Officer Data observes, "the life form is draining energy directly from the fusion reactors." Counselor Troi notes, "It's feeding off the energy of the *Enterprise* as it would from its mother."

Life forms need energy for maintaining themselves and eventually for reproducing. Energy must be captured from the environment, biochemically stored, and processed to be made available for use when needed. Animals such as humans consume other life forms (plants, fungi, and other animals) to get sugars and other nutrients to use as stored biochemical energy. We then use respiration to metabolize sugars and produce molecules that make energy more accessible to the cell (e.g., adenosine triphosphate, or ATP[§§]). Respiration can involve breaking down sugar in the presence of oxygen (called "aerobic respiration") to make ATP, also producing carbon dioxide and water as side products. Much of this process occurs in cellular structures called mitochondria, to which we will return in chapter 2. ATP is the immediate source of energy for many of life's processes, including transportation of molecules in and out of individual cells, cell division, muscle movement, and many more.

Similar processes lead to the breakdown of sugars and production of accessible energy without needing oxygen. Some bacteria and archaea break down sugars to produce energy without oxygen (via "anaerobic respiration"), particularly species that live in low oxygen environments. One example is the archaea

§§   The holographic doctor mentions adenosine triphosphate in regard to the captain's respiration in *VOY* "Sacred Ground."

that live in the gut of cows and other ruminants; these species help break down cellulose from plant material into energy, incidentally producing methane gas as a by-product. Oxygen is actually detrimental to the growth of some such organisms. Another process called fermentation can produce ATP from the breakdown of sugars without the need for oxygen gas. This process happens in our own muscle cells at times when we have too little oxygen for standard aerobic respiration, with an additional product being lactic acid. Many people are more familiar with the use of fermentation by yeasts, though, wherein both ATP and ethyl alcohol are products. In general, however, these processes tend to produce less energy than aerobic respiration in the organisms we know.

All of the above assumes that accessible energy for the cell (e.g., ATP) is primarily formed from the breakdown of sugars or other molecules acting as biochemical energy stores. In some sense, this pushes the question of "energy source" back a step: how are these sugars formed? Many readers will be at least loosely familiar with the process of photosynthesis, whereby plant cells take up carbon dioxide and water from the environment and convert them to simple sugars and oxygen using energy derived from sunlight. This process is common in plants, algae, and some forms of bacteria, and they absorb the energy from light in this process using a pigment, the most well-known being chlorophyll. Some other species also photosynthesize sugars but do not use water or produce oxygen as a result. For example, purple sulfur bacteria use light energy to convert carbon dioxide and hydrogen sulfide to sugars and elemental sulfur. In this case, photosynthesis works best under low light intensity and without oxygen, so these bacteria are found in salt marshes, in marine environments, and sometimes even in the guts of marine zooplankton.[37]

Although sunlight coming to the Earth's surface may be an obvious energy source for sugar production, it is not the only energy source even among life forms on Earth. Some bacteria that live in deep parts of the ocean or in undersea hot springs use "chemosynthesis" rather than photosynthesis to make sugars, not requiring sunlight but instead utilizing energy released by chemical reactions involving hydrogen sulfide or methane. One species of green sulfur bacteria photosynthesizes using geothermal radiation rather than sunlight.[38] Finally, some organisms seem to use ionizing radiation (e.g., gamma rays or X-rays) as an energy source. For example, at least two fungal species grow faster when exposed to ionizing radiation,[39] and the bacterium *Desulforudis audaxviator* gets energy to make sugars from the decay of radioactive uranium in underground mines.[40] This last example opens the question of whether life could, in principle, be powered by radiation from galactic cosmic rays hitting planetary surfaces.[41] Such rays may be a powerful source of energy for life on planets bearing thin atmospheres.

I have stressed just some of the diversity of forms of energy production or release throughout this section but still only scratched the surface of the dimensions of diversity here on Earth. Again, we predict that life in outer space (and in *Star Trek*) should be at least as diverse, and arguably more diverse, than life on Earth. Basic animal and plant life are observed regularly through much of *Star Trek*, though typically the latter is not emphasized (see first example in Appendix). However, the silicon-based species mentioned at the beginning of this section on the planet Velara III in *TNG* "Home Soil" was said to be "photoelectric," gaining energy from light. The crystalline entity in *TNG* "Datalore" and "Silicon Avatar" converts organic matter to usable energy, leaving behind hydrocarbons; as the organic matter in each case existed in an oxygen atmosphere, this could be analogous to the energy-producing reactions involved in terrestrial respiration (if we ignore the problems raised in the previous section about silicon). As for energy producers, the *Enterprise* crew found a planet with plant life nowhere near a star in *ENT* "Rogue Planet"— perhaps the inhabitants obtained energy from the subsurface "thermal vents" mentioned in the episode.

In terms of energy usage, the *Enterprise-D* encounters space-dwelling beings that can drain energy from its fusion reactors in *TNG* "Galaxy's Child" (quoted at the beginning of this section). What puzzled me in this episode was how an individual of this species drained energy without directly plugging into the reactors; the implication was that it was taking energy that would have been used by the ship rather than merely absorbing energy that was being lost into space. As an example of this point, photosynthesis by plants or energy absorbed by solar panels does not "drain" the sun in some way. However, the space-dwelling being was attached to the ship's outer hull, so perhaps it was somehow drawing the power via conduits. Droplet-sized organisms called GS54 also drained energy from a spaceship from its exterior in *DIS* "Context Is for Kings." Assuming some direct connections, such a drain may be possible in principle: some Earth microbes can directly accept and use electrons from electrodes for energy.[42] In any case, I applaud the writers for thinking of unusual forms of energy production or usage by living forms.

Relatedly, *Star Trek* does mention a few extraterrestrial species that do not breathe oxygen, and presumably do not respire as we do. The Axanar in *ENT* "Fight or Flight" breathed a nitrogen-methane combination, and the Tesnians in *ENT* "Shuttlepod One" required "boron gas" (perhaps referring to boron bound to hydrogen, fluorine, or chlorine, since elemental boron is a solid at room temperature and pressure). Young Lorillians in *ENT* "Broken Bow" were said to breathe the gas "methyl oxide" (a compound that either does not exist or is imprecisely named by our present standards: either two methyl groups would bind to oxygen resulting in what we call "dimethyl ether," or a double bond of oxygen to CH_2 would produce what we call formaldehyde or methanal). *DS9* also had a few species with distinct breathing requirements: the Yalosians (60% nitrogen, 10% benzene, and 30% hydrogen fluoride, in the episode "Improbable Cause"), the Lothra (hydrogen, in "Melora"), and possibly an unidentified silicate-based shape-shifting species (carbon dioxide at high concentrations, in "The Alternate"). There was a passing reference to anaerobic bacteria on the planet Kataan in *TNG* "The Inner Light." Finally, of course, each of the series depicted various species that survived in space without ships or suits, so clearly not requiring (frequent access to) oxygen. Overall, the *Star Trek* series incorporate some diversity of approaches to energy production and energy processing for life.

CLOSING REMARKS

Isaac Asimov complained in 1962 of "the lack of imagination in movieland's monsters. Their only attributes are their bigness and destructiveness. They include big apes, big octopuses (or is the word 'octopodes'?), big eagles, big spiders, big amoebae."[43] *Star Trek* is not entirely innocent of Asimov's complaint, in that one has to dig into the 6 nonanimated series and 13 movies to find references to what appear to be entirely "new" life forms, but I would suggest the series arguably do better than many popular science fiction depictions on television. I have only touched on a few parameters of life in this chapter, and I have left aside many obvious ones (e.g., extremes of size, or, as in *TNG* "Interface," living while floating in a gaseous atmosphere rather than on solid surfaces or in liquids).[44] Further, much of this chapter has focused on attributes singly rather than in combinations. For an organism to survive on parts of Mars, for instance, it would have to be tolerant simultaneously of cold temperatures (especially at night), very low atmospheric pressure, lighter gravity,[***] high radiation (via solar radiation and cosmic rays), virtual absence of liquid water and gaseous oxygen, and other demands. Most extraterrestrial life would very likely need to be simultaneously tolerant of multiple conditions that we perceive as extreme on Earth.

One other dimension worthy of more consideration in extraterrestrial life is "speed." We think of life operating in time scales with which we are familiar. Could life exist at a much, much faster speed that makes it virtually unobservable? *Star Trek* touched on this briefly in *TOS* "Wink of an Eye," with humanoids that moved around them so quickly that they sounded like a mere buzz to the crew, and *VOY* "Blink of an Eye," in which time on a planet moved faster than in surrounding space. However, in both of these cases, the fast speed seemed a by-product of the environment rather than a fundamental difference in the biological processes of the life forms present. Alternatively, chemical or other reactions associated with life may proceed on a glacially slow time scale. Again, this came up briefly in *TNG* "Tin Man" with reference to the slow Chandrans, who took three days to say hello. However, one can imagine the possibility of life on a much slower time scale yet, potentially requiring millennia of observation to document basic life functions (as can be true for some Earth psychrophiles). Many other dimensions are also worth considering, going beyond what we find familiar.

[***]   One example of this from *Star Trek* is the Elaysian in *DS9* "Melora," whose physiology was not adapted to Earthlike gravity; see chapter 5.

Generally speaking, the mission of "seeking out new life" both in science and in science fiction can benefit from greater imagination regarding what "life" may be, its chemical and energetic bases, and where and how it may be observed. Still, the *Star Trek* writers have provided approachable entertainment that occasionally taps into these questions. They do "boldly go" at least a little way beyond where anyone has gone before (in real life). In subsequent chapters, I will delve less into diversity and more into the evolutionary biology of organisms both on Earth and as depicted in the series.

CHAPTER 2

CHARTING THE RELATIONSHIPS OF SPECIES

There is grandeur in this view of life, with its
several powers, having been originally breathed
into a few forms or into one; ... from so simple a
beginning endless forms most beautiful and most
wonderful have been, and are being, evolved.

—CHARLES DARWIN, 1861[1]

Darwin is credited with one of the most fundamental advances
in our understanding of life: that all life forms on Earth (the
"endless forms" mentioned above) are directly and hierarchically
related to each other, and descended from a single shared ancestor (the "one" mentioned above). In this sense, all life on Earth
forms a single family tree spanning roughly four billion years.
On a broader scale, we humans are like brothers with other primates, such as orangutans and gorillas; cousins of other verte-

brates, including fish, frogs, reptiles, and birds; second cousins of other animals, for example, worms, flies, crabs; and distant relatives of plants and fungi. We are very distantly related even to archaea and bacteria.

With this concept of "shared ancestry" in mind, this chapter explores evidence for the relatedness of species on Earth as well as misconceptions of such relations and misuse of terms like "higher" and "lower" forms of life. The chapter concludes with a detailed exposition of three hypotheses from *Star Trek* regarding how fictional humanoid life on other worlds may be related to life on Earth. The three hypotheses are assessed using evolutionary principles and approaches discussed earlier in the chapter. While I apply the tests to fictional humanoids from the *Star Trek* series, the same principles and logic can be applied, for example, to inferring the origin of a life form on Mars resembling an Earth microbe.

EVIDENCE AND MISCONCEPTIONS OF COMMON ANCESTRY

EVIDENCE

DS9, Season 1, Episode 20, "In the Hands of the Prophets"
Keiko O'Brien is teaching a group of schoolchildren on the space station about a stellar phenomenon located near the planet Bajor. Vedek (a clerical title) Winn Adami from Bajor observes the class and expresses concern that Mrs. O'Brien does not present this phenomenon as containing the "Celestial Temple of the Prophets" as per Bajoran spiritual beliefs. Mrs. O'Brien instead discusses it in purely scientific terms. Vedek Winn argues that this teaching is blasphemous. Later,

Vedek Winn proposes a compromise that Mrs. O'Brien just
not discuss the phenomenon at all. Mrs. O'Brien retorts,
"And when we get to theories of evolution or creation of the
universe, what then?"

Despite low levels of public "belief" in evolution in some parts
of the world, such as the United States and the Middle East (and
potentially on the planet Bajor in *DS9*), scientific consensus and
overwhelming evidence support the shared ancestry of all life
on Earth. This evidence proves that all present-day species descended from earlier species, and that we share ancestors with every modern-day species on our planet. I will delve in some detail
into some of the scientific evidence for common ancestry below,
and there are other excellent resources for a deeper exposition of
this topic (see Suggested Associated Scientific Reading).

Species on Earth share many general attributes that demonstrate they are directly related. As discussed in chapter 1, all
known life forms are water and carbon based. The nucleic acids
DNA and RNA mediate inheritance in all life on Earth. Inside
living cells, DNA and RNA sequence codes (made up of varying
combinations of building blocks called nucleotides) guide the
production of specific amino acids that ultimately combine to
make proteins. Despite the large number of amino acids—20 in
most species—most of the specific DNA and RNA codes used
to guide the addition of each amino acid to a protein are exactly
the same in all life forms. For example, exactly the same three-
nucleotide DNA sequence encodes the amino acid methionine
in humans, mosquitoes, sunflowers, baker's yeast, and *E. coli* bacteria. Such striking similarity in genetic code would be highly
improbable if species arose independently: there is no a priori
reason for the association of that particular three-nucleotide
DNA sequence with the amino acid methionine. One recent

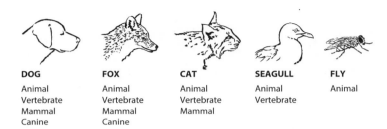

DOG   **FOX**   **CAT**   **SEAGULL**   **FLY**

Animal   Animal   Animal   Animal   Animal
Vertebrate   Vertebrate   Vertebrate   Vertebrate
Mammal   Mammal   Mammal
Canine   Canine

Figure 1. Hierarchical classification of example animals.

study ran an elegant statistical analysis on amino acid sequences from across the major divisions of life and found exceedingly high support for universal common ancestry,[2] though other work questioned some details of the analysis.[3]

Many of us learned the hierarchical classification of life in grade school. This classification system was based on one originally introduced in 1735 by Swedish biologist Carl Linnaeus in *Systema Naturae*.[4] Let us explore an abbreviated version of classification with dogs as an example (see figure 1). In hierarchical order, dogs are animals, vertebrates (having a backbone), mammals (being warm-blooded, having hairy bodies, and producing milk to feed offspring), and canines (sharing aspects of appearance and behavior). Foxes fit within exactly the same hierarchy, and therefore are very similar to dogs. Cats are animals, vertebrates, and mammals, but not canines—they are similar to dogs, but less similar to dogs than foxes are. Seagulls are animals and vertebrates but they are birds and not mammals. Fruit flies are animals but not vertebrates or any of the other classifications noted above. In that sense, all five of these creatures are animals, four (dog, fox, cat, bird) are vertebrates, three (dog, fox, cat) are mammals, and two (dog, fox) are canines. Using such a classification system, we can define which animals (or organisms

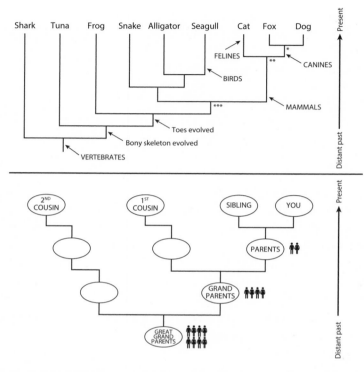

Figure 2. Inferred relationships among species (*top*) and among individuals in a family (*bottom*). See main text for an explanation of the asterisks.

more generally) are most similar to each other in a hierarchical manner.

The modern biological view is that these hierarchical relationships reflect ancestry (figure 2). We can also observe an abbreviated version of these similarity hierarchies within human families. On average, sisters look more similar to each other than they do to their first cousins. First cousins look more similar to each other on average than do second cousins. Hence, if one were to classify members of a particular generation into "groupings" solely by appearance and characteristics, siblings would share the most traits (like dog-fox above), cousins would share some traits (like dog-cat above), and people from the same part of the world would share a bit more than people at random (like dog-bird versus dog-fly above).

Continuing this analogy, we expect the most similar groups to share the most recent ancestors. Siblings share parents (one generation back), cousins share grandparents (two generations back), etc. The less related two individuals are, the further back in time it will be until you find the shared ancestors. The same holds with species. Dog and fox share a recent ancestor (depicted as * in figure 2, not far down from present at the top row), dog and cat share an ancestor further back in time (**), dog and bird share one still further back (***), and dog and fly share an ancestor very long ago (not shown).

One of the most striking lines of support for common ancestry as the best explanation for the hierarchical relationships is that the same hierarchy emerges even when very different kinds of traits are studied. On seeing it for the first time, one can identify a species as a "bird" not just by it having wings but also by it having feathers, a beak, etc. Traits are not randomly distributed across species or groups. Classification to groups can also be performed using DNA sequences as traits. Similarity in DNA sequence implies close relationship, and the hierarchical relations inferred from DNA match almost identically with those from visible features. Figure 3 depicts a specific gene's DNA sequence as identical between dog and fox, differing at two sites between dog and cat, differing at three sites between dog and seagull, and differing at six sites between dog and fruit fly. This convergence of evidence for the same hierarchy among diverse traits and genes provides strong support for the relatedness and common ancestry of life on Earth.

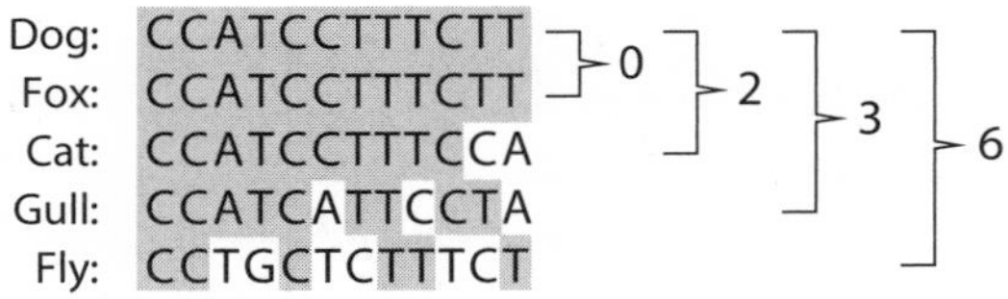

```
Dog:   CCATCCTTTCTT ⎫
Fox:   CCATCCTTTCTT ⎬ 0 ⎫
Cat:   CCATCCTTTCCA      ⎬ 2 ⎫
Gull:  CCATCATTCCTA          ⎬ 3 ⎫
Fly:   CCTGCTCTTTCT              ⎬ 6
```

Figure 3. Partial DNA sequence from the cytochrome c oxidase subunit I gene from five species, with the number of nucleotide differences from the dog sequence highlighted. Incidentally, this gene produces part of an enzyme used in aerobic metabolism in all of these species.

While many characteristics suggest the same hierarchical relationships, occasionally some characteristics do not match perfectly. For instance, figure 2 illustrates the origin of toes, but snakes have no toes (or legs at all), despite descending from toed ancestors. Nonetheless, in such cases, many other traits support the relationships depicted, and the "weight of evidence" is strongly in favor of a particular relationship. Snakes, for example, share physical features (e.g., dry, scaly skin) and DNA sequences with other reptiles, and abundant evidence shows that their ancestors had limbs but subsequently lost them. Indeed, some snakes, such as pythons and boa constrictors, still have tiny leg bones buried under their skin.[5]

Like the snake leg bones mentioned, many species retain traits that had a specific function in their ancestors but no longer serve that function—"vestigial" traits. Ostriches have vestigial wings, for example, that are not able to help them fly. However, since ostriches descended from birds that could fly, it makes sense that they might have wings. Genes can also be vestigial. Most mammals have a gene that produces vitamin C from a simple sugar in their bodies. Many primate species (including humans) inherited a mutation in this gene that inactivates it, and this mutation requires us to get vitamin C from our diet. However, the inactive gene is still present in the DNA of our cells and, indeed, some function of the gene can even be restored via genetic manipulation.[6] The presence of such vestigial traits (and genes) only makes sense in the context of the relatedness and common ancestry of species, and it also demonstrates that evolution does not always lead to greater complexity.

Finally, the most famous evidence for common ancestry is the fossil record buried on Earth. Fossils are remains or imprints of long-dead forms. After being buried by sediments, some kinds of remains—such as bones or shells—are gradually replaced by minerals, and some remains leave behind imprints or residues that are preserved in hardened sediments. We can infer how long ago the fossilized organisms lived from the age of the rocks in which they are present. These fossils sometimes depict transitional forms between many present-day species or groups, confirming (or sometimes tweaking) our perceptions of the relationships between them. For instance, several fossil snakes bearing legs have been discovered: some are possible ancestors of modern snakes, or near relatives of such ancestors, which predate the complete loss of external limbs in modern snakes.[7] Many transitional forms have recently been discovered linking modern birds to their theropod reptilian ancestors.[8] Similarly, researchers have found compelling links between ancient hoofed animals and modern-day cetaceans, such as humpback whales, through studies of fossils and in comparison to present-day species.[9] (As an aside, I hope that the movie *Star Trek IV: The Voyage Home* is wrong about humpback whales going extinct in the twenty-first century—recent data suggest that many populations are anticipated to be stable for the coming 60 years, barring any major change.)[10]

The *Star Trek* series admirably recognize evidence for evolution and common ancestry of species on Earth. Doctor Ann Mulhall in *TOS* episode "Return to Tomorrow" notes that, "our studies

indicate that life on our planet, Earth, evolved independently."
The holographic doctor in *VOY* mentioned a shared ancestor of
cold-blooded and warm-blooded species on Earth in "Distant
Origin." (I will return to specifics from these two episodes later
in this chapter.) Lieutenant Commander Data in *TNG* "Shades
of Gray" notes, "True fossilization requires several millennia,"
presumably referring to replacement of organic remains with
minerals and the formation of stone imprints. The series finale
of *TNG* "All Good Things . . ." had Captain Picard transported
back to the chemical origin of life on Earth, explicitly acknowledging that all subsequent life on Earth descended from a single
microscopic-scale incident.

Despite all of this overwhelming scientific evidence, and its
general acceptance in the *Star Trek* series, common ancestry of
species is still controversial in many parts of the present world.
Further, even though the *Star Trek* series present a hopeful vision of our scientific future, the quote at the beginning of this
section implies that some controversy persists about teaching
evolution even in the fictional twenty-fourth century. Some of
this doubt comes from misunderstandings and misconceptions,
and the next section elaborates a few of the more common ones.

MISCONCEPTIONS

TNG, Season 7, Episode 19, "Genesis"

Captain Picard and Lieutenant Commander Data return
to the *USS Enterprise-D* and find that all the crew have
been infected with a virus that has caused them to "de-
evolve" into other forms. One crew member has become
an amphibian, others nonhuman primates, and one has
become something akin to a spider. Data's pet cat has de-
evolved into an iguana. Data explains to the captain, "Each

of these stages is another link in the evolutionary chain which stretches back to the origins of all life forms on Earth. . . . They are each de-evolving to earlier forms of life."

Common ancestry of all life has numerous implications for understanding life on our planet, and sometimes misconceptions arise from inadequately considering these implications. First, evolution has occurred over the same total amount of time in all modern species: we all evolved over almost four billion years— since the origin of life on Earth. This time has an important consequence: while humans evolved within this time period, so too has evolution happened in every other lineage: bacteria, archaea, plants, fungi, and other animals. Even organisms that may appear superficially similar to some of their ancestors (such as some microbes) have evolved over this same time, potentially dramatically in some ways.

In this regard, common expressions such as "more evolved" or "higher" forms of life are imprecise.* Conceivably, one could try to quantify the amount of change over time, but it becomes unclear what the unit of measurement should be: physical size, number of DNA nucleotides changed, time required to solve a maze, or something else? We think of ourselves as "higher" life forms, but we cannot photosynthesize like a plant, survive submersion into liquid nitrogen like a tardigrade, or withstand strong ionizing radiation like the microbe *Deinococcus radiodurans*. In the *TOS* episode "Errand of Mercy," Mr. Spock notes of the alien Organians that they are "as far above us on the evolutionary scale as we are above the amoeba." While we certainly have many abilities that amoebas do not share, amoebas are shape-shifters and can produce hardened cysts around their

---

* Another reference like this was made in *VOY* "Nothing Human."

bodies to withstand stressful conditions—advantageous traits we do not have.

Relatedly, we did not evolve "from" other present-day life forms. We did not evolve from a modern chimpanzee, a modern seagull, or a modern tuna. However, we share ancestors in common with each of those other modern species. Just as there was evolutionary change from those ancestors leading to the modern human, there was also evolutionary change from those ancestors leading to the modern chimpanzee, seagull, or tuna. If we use the analogy of the family tree, all modern species are in the same "present" generation (the top row of the family tree in figure 2). We did not evolve from these other species any more than we are descendants of our siblings or cousins.

In this regard, the incidents described in the *TNG* episode "Genesis" above are implausible in several respects. If humans were to revert to ancestral forms, a primitive primate or an ancient amphibian are indeed possible (setting aside the implausibility of the proposed mechanism until the next chapter), but we have no ancestors that were spiders. Spiders evolved from ancient arthropods, but the first spiders arose in the arthropod lineage long after the split from the lineage leading to vertebrates. To continue the analogy above, our species is related to ancient spiders like people are related to their mothers' cousins—we share ancestors with them but we are not direct descendants from them. For this same reason, Data's cat may have had a vaguely reptilian ancestor, but presumably no ancestor of cats looked like a modern iguana. Hence, Data's statement that they are evolving to "earlier" forms of life is misleading. Still, I commend the series for acknowledging the direct relationships of life forms on Earth even if they slipped on the specifics of how these relationships operate.

Impressively, the *Star Trek* series brought forward the question of why "humanoid" aliens exist on other worlds, beyond the obvious answer of simplicity with respect to use of human actors in a television series. Many science fiction series feature alien forms superficially resembling humans or other Earth-bound life forms, but I applaud *Star Trek* for addressing the similarity directly. One simple, though astronomically improbable, answer is that they evolved completely independently. Some convergence of form may be predicted via natural selection, and I return to this possibility in chapter 4. However, I discuss and assess below some more reasonable alternative hypotheses presented in the series for why humanoid aliens might exist.

Importantly, the predictions presented below and used for assessing the three hypotheses regarding extraterrestrial humanoid life are not unique to examining "humanoid" life in particular. If a probe on Mars or Europa were to find microbes similar to Earth's bacteria, similar hypotheses might be posed, and the same approaches to assessing them might be taken. Hence, this exercise does more than merely analyze the *Star Trek* series; it suggests approaches for the study of any extraterrestrial life resembling life that we know on Earth.

RELATIONSHIP OF LIFE ON OTHER WORLDS

HYPOTHESIS 1: DIRECTED PSEUDOPANSPERMIA

TNG, Season 6, Episode 20, "The Chase"

A computer program generated four billion years ago but encoded in the DNA of a large number of species produces a hologram that speaks to the species in attendance—humans, Klingons, Romulans, and Cardassians. It describes its species' history of exploring space but not finding other similar life forms. "So, we left you. Our scientists

seeded the primordial oceans of many worlds, where life
was in its infancy. The seed codes directed your evolution
toward a physical form resembling ours.... The seed codes
also contained this message, which we scattered in fragments on many different worlds."

We can derive three predictions for what would need to have
happened in the evolutionary history of life for the hologram's
words to explain the similarity of humanoid life across planets.
First, one might interpret the monologue as suggesting that at
least some of the raw materials for life on Earth and on other
worlds came from outer space rather than life arising completely
in situ.[†] Second, a "seed code" has persisted within life on Earth
(and on other worlds) for the past four billion years. Finally,
that a persisting seed code predictably directed the evolution of
humanoid life on Earth and on other planets in the same four
billion years. I explore each of these predictions in turn.

The first prediction regarding extraterrestrial origins of Earth
life has been discussed for many years under the general heading
of "panspermia."[‡] In its most basic form, the panspermia explanation posits that one or more already living (likely microscopic)
forms were introduced to Earth from outer space, and these forms
subsequently evolved into the diversity of life we have on Earth
today. A variant of the explanation, called pseudopanspermia or
soft panspermia, suggests the organic molecules that spurred or
continued the origin of life on Earth came from space.[§]

† The quote is cryptic with respect to what exactly constituted life "in its infancy" on Earth—an incomplete (or unassembled) set of raw materials for life
or some actual living forms.
‡ This term was also used in *DIS* "Context Is for Kings" in relation to spores
spread through space.
§ This version seems consistent with the conversation between Q and
Capt. Picard in later episode *TNG* "All Good Things ... ," when they went back
in time to the origin of life on Earth. Q said, "Right here, life is about to form on

Pseudopanspermia-based explanations for life on Earth are quite reasonable. As mentioned in the preceding chapter, scientists have isolated extraterrestrially formed amino acids and components of DNA and RNA from meteorites.[11] The amino acid glycine has been found in samples taken from comet 81P/Wild 2.[12] Clearly, some raw materials associated with life on Earth have formed outside our planet and sometimes have been introduced to our planet from extraterrestrial sources.** The introduction of a living being to Earth (the basic form of panspermia) is somewhat more challenging, since such a form would need to survive the transit to Earth. One possible means of survival, however, is that the life form is a resilient microbe inside a large rock that was launched into space via an impact. Some studies have shown that bacterial or fungal spores could potentially survive some aspects of such a transfer.[13]

The second prediction, regarding persistence of seed codes over billions of years on Earth (and other worlds), is problematic. To assess the prediction, I will assume that the seed codes exist within some portion of the nucleotide sequence of human DNA. With several assumptions, we can calculate the probability that a nucleotide sequence of DNA would remain unaltered over billions of years. Rather than using all 4 billion, I will focus on just the first 1.5 billion years, when our ancestors were all microbes. Let us assume that (1) these nucleotides serve no other function than to produce a holographic message, (2) the rate of random mutation in the code is roughly one in 10 billion per nucleotide per generation,[14] and (3) this early life goes through roughly

this planet for the very first time. A group of amino acids are about to combine and form the first protein, the building blocks of what you call life."

** *VOY* "Year of Hell" also briefly mentioned hydrocarbon fragments from a comet impacting a planet in the Delta Quadrant four billion years ago giving rise to (plant) life there.

100 generations per year (or one cell division per 3.6 days). With those assumptions in mind, the probability that one specific nucleotide would be transmitted unaltered over 1.5 billion years is equal to $0.9999999999^{150,000,000,000}$, or roughly 1 in 3 million. For context, this is about half the probability of the reader of this chapter being killed by a falling meteorite, asteroid, or comet.[15] Further, the elements of the message would likely need to be encoded across many nucleotides rather than as just one. In short, such a message would almost certainly have degraded to incoherency.

One can potentially question some of my assumptions above. For instance, perhaps the same DNA code containing the message also confers an essential function (e.g., it is associated with the process of making ATP) and therefore is more likely to persist unaltered in living cells. Alternatively, the message may not be in the DNA's nucleotide sequence itself but instead exists as something more fundamental about the molecular structure of DNA (e.g., the kinds of molecules in nucleotides). Other parts of the *TNG* episode are unspecific (and muddled) about the nature of the persistent code, including references to DNA strand fragments, uniform base pair combinations, "protein-link compatibility," etc., so alternatives to nucleotide sequence are difficult to assess with the information given.

Finally, even if we were to assume such a code persisted, the inference that humans (or even a general humanoid form) were a predetermined outcome of these precursors four billion years earlier is exceedingly unlikely. Paleontologist Stephen J. Gould famously proposed a thought experiment wherein one would "replay life's tape": essentially, if we were to rewind life to where it was hundreds of millions of years ago and then let things play out again, would we still have the same outcome?[16] The implication from the *TNG* episode described above is that humanoids would emerge again—and did emerge multiple times elsewhere. However, this implication suggests that "random" (unpredictable) events played little role in shaping overall humanoid form billions of years later. In contrast, Gould concludes that many events leading to the current diversity on our planet (including mammals in general and humans in particular) were indeed random. For example, the volcanic activity and asteroid impact that caused the mass dinosaur extinction about 66 million years ago were necessary for mammals to emerge and diversify.[††] Had these random events not occurred, large reptiles would have dominated the planet for a longer time and delayed or prevented the evolution of large mammals. As Gould notes regarding the asteroid impact, "In an entirely literal sense, we owe our existence, as large and reasoning mammals, to our lucky stars."[17]

There are many, many more well-known examples of historical accidents that led directly to modern life (including humans). The popular "endosymbiotic theory" proposes that mitochondria, the powerhouses of animal, plant, and fungal cells, arose when an ancient single-celled organism engulfed (swallowed) a small bacterium that utilized oxygen and produced energy via aerobic respiration (see chapter 1).[‡‡] The engulfed bacterium was not broken down but instead reproduced within the cell, and when the cell divided, the daughter cells each had some of the bacteria.[18] Eventually, the intracellular bacteria evolved into the organelles we now call mitochondria, and the abundant energy that mitochondria provide host cells likely facilitated evolution of the multicellular organisms we know.[19]

††   This event was also mentioned in *ENT* "Azati Prime."
‡‡   This is also sometimes called "symbiogenesis" and is mentioned briefly and indirectly in *VOY* "Tuvix" as organisms of different species merging to form a third, unique species: see Appendix.

Quite admirably, endosymbiosis was even described briefly in a *Star Trek* episode: *VOY* "Once Upon a Time." This rather random event may thus have completely changed the course of evolution on our planet, both in general and with respect to animals (including humans) in particular. Given these events, humanoids were almost certainly not a predestined outcome of evolution.

Some biologists still argue of the relative importance of chance versus determinism in some specific cases of evolutionary change (a topic to which I return briefly in chapter 4), but no scientific principles support beings so perfectly humanoid evolving repeatedly over billions of years. While I applaud the episode for introducing pseudopanspermia as a possible means for the origin of life on Earth, the directed, predetermined evolution of humanoids following such an origin is exceedingly unlikely. Fortunately, the *Star Trek* series offer alternative hypotheses.

HYPOTHESIS 2: ANCIENT ALIEN ASTRONAUTS

TOS, Season 2, Episode 20, "Return To Tomorrow"

Several members of the *Enterprise* crew are in an underground chamber, and they come upon a glowing ball on a pedestal. The ball speaks to them, identifying itself as Sargon. It claims to be the essence of a mind sealed in a receptacle, but notes it once had a body similar to the crews? It also notes that humans may be the descendants of its species. "Six thousand centuries ago, our vessels were colonizing this galaxy, just as your own starships have now begun to explore that vastness. As you now leave your own seed on distant planets, so we left our seed behind us. Perhaps your own legends of an Adam and an Eve were two of our travelers."

Star Trek was not the first to suggest that ancient aliens came to Earth and either influenced existing humans or are the ancestors of modern humans. Such ideas permeated science fiction both much earlier[20] and later—for example, the 1970s science fiction TV series *Battlestar Galactica* opened with, "There are those who believe that life here began out there," suggested that aliens came to Earth, and concluded with the idea of "brothers of man" existing on other worlds. Arguments used to support an ancient alien presence on Earth stem from legends of unexpected feats that past civilizations accomplished or examinations of persisting archaeology (e.g., the pyramids of Egypt). American astronomer Carl Sagan examined "the long litany of 'ancient astronaut' pop archaeology" in multiple books and concluded "there are always more plausible alternative explanations based on known human abilities and behavior."[21]

While we cannot exclude the possibility of ancient alien astronauts *visiting* Earth, we can assess the hypothesis that humans have alien ancestors dating to roughly 600,000 years ago as suggested in the episode. First, the hypothesis predicts that humanoids were present on Earth roughly that long ago. Second, evidence should exist that modern humans descend from these spacefarers and not from other organisms found on Earth before that time, thus being unrelated to other life forms on Earth.

The first prediction is straightforward: abundant evidence suggests that multiple humanoid species were on Earth roughly 600,000 years ago. For example, that timing coincides with evidence in the fossil record of *Homo erectus*, forms similar to which had spread across much of southern Asia and parts of Africa. These people are presumed to have been hunter-gatherers who used tools and ate a diverse diet ranging from nuts to elephants.[22] Some evidence suggests that groups of these people used fire, and more extensive use of fire is known to have

occurred 300,000–400,000 years ago.[23] However, while humanoids undoubtedly roamed the Earth 600,000 years ago, evidence
of humanoids is also present much earlier, so the episode perhaps implies that "modern" humans descended from aliens
while these ancient forms went extinct. This implication leads to
the second prediction.

Can we test the prediction that ancient alien astronauts are
modern humans' forefathers, perhaps even allowing a few million years of flexibility to the exact timing? Few fossils exist of
species that are likely to be (or are closely related to) our most
recent shared ancestors with chimpanzees,[24] so researchers have
sometimes tried to infer the form of such ancestors using comparative methods.[25] One might use the potential gap in the
known fossil record to propose that the human lineage could
have arisen independently of other primates and/or have extraterrestrial origins, improbable as such convergence in form may
seem given the structural similarity of forms present in Earth's
fossil record and alive today. However, such a (preposterous) explanation ignores the extensive lines of evidence for common
ancestry described in the beginning of this chapter, all of which
apply as readily to humans as to other organisms. For example,
humans differ from chimpanzees at only ~1% of DNA nucleotide sites (see chapter 3).[26] Chimpanzees differ from orangutans
or gibbons at more nucleotide sites than they do humans. If
humans arose independently (extraterrestrially or otherwise),
then there is no reason for our DNA sequences to be so similar
to those of chimpanzees. As such, the hypothesis proposed in
the episode is not supported: our origins trace to Earth, we are
closely related to chimpanzees, and the divergence of our ancestors from chimpanzees occurred well over 600,000 years ago.

To the credit of the *Star Trek* writers, this hypothesis was
quickly downplayed, even within the episode mentioned above.

Immediately after Sargon's statement, astrobiologist Doctor Ann Mulhall noted, "Our beliefs and our studies indicate that life on our planet, Earth, evolved independently."[§§] While I am uncertain why "beliefs" was used in the statement—implying religion or philosophy rather than science—the scientific consensus is certainly that humans evolved from primate ancestors on Earth. A variant of this hypothesis was proposed in *VOY* "Tattoo," where aliens were proposed to have visited Native American humans on Earth and given them a "genetic gift" of some sort that affected their migratory behaviors. This hypothesis is not falsifiable by itself, at least without more information. However, while it explains how some humans may have been influenced by aliens, it does not explain our main question of why humanoid life like that on Earth is found elsewhere. On this main point, I now move on to the third hypothesis.

HYPOTHESIS 3: FROM EARTH TO THE STARS

TOS, Season 3, Episode 3, "The Paradise Syndrome"

The *Enterprise* crew come across a population of Native American humans on a planet, living near an alien obelisk. Spock eventually translates symbols on the obelisk to determine its origin. He explains, "The obelisk is a marker. . . . It was left by a super-race known as the Preservers. They passed through the galaxy rescuing primitive cultures which were in danger of extinction and seeding them, so to speak, where they could live and grow." Apparently, the Preservers seeded multiple humanoids scattered throughout the galaxy.

§§   *TOS* "Return to Tomorrow."

This hypothesis is essentially the opposite of the preceding one, placing the origin of extraterrestrial humanoids on Earth rather than suggesting the origin of Earth-dwelling humans stems from some other planet. Together with a discussion early in the episode regarding plant life, this episode seems to imply that much of the ecosystem (humans, plants, and presumably microbial, fungal, and other animal life—for example, "fish nets" were mentioned at one point in the episode) was transplanted.

The hypothesis leads to two overlapping predictions. First, life originating on Earth has been introduced to other worlds. Second, humanoid forms found on other worlds in the series were "recently" transplanted to their current home planets from Earth. Unlike the predictions from the two preceding hypotheses, the first of these predictions is very likely true, and the second would be easily testable if such alien humanoid species existed.

Life forms from Earth may well have been introduced to Mars and other worlds, but they are far more likely to be microbial than humanoid. Some of these introductions could have occurred through processes not involving humans: impacts of asteroids or large meteors with Earth over the past four billion years may have ejected chunks of Earth's crust (likely bearing microbes) that landed on Mars.[27] Several meteorites from Mars have been found on Earth, and while the reverse direction may be less likely, the possibility exists.[28] Additionally, humans have sent spacecraft to Mars that bore microbial spores. For instance, the Mars *Curiosity* rover had hundreds of thousands of spores on its flight system prior to launch.[29] The Committee on Space Research Panel on Planetary Protection provides specific guidelines for the level of sterilization needed for spacecraft, depending on their destination, but none of these guidelines insist upon the (unfeasible) goal of "complete sterilization" of all components of a spacecraft.[30] With either of these means of transit, natural or via human spacecraft, the microbes would need to survive the transit to another world—including ejection from Earth, time in space, and the crash landing on the alien world—and survive under the conditions of that other world. They would need to survive intensely high and low temperatures (by Earth's-surface standards), pressure changes, dehydration, and radiation. Clearly, most organisms (microbial or otherwise) would die from these extreme conditions, but a subset may survive as spores. Chapter 1 discussed some microbes exhibiting extreme tolerance to many of these factors as well as the ability to remain metabolically inactive for thousands of years. Such organisms are more likely yet to survive the journey if shielded en route either within a chunk of Earth's crust or inside a spacecraft component. We do not yet have data demonstrating whether or how many organisms have survived a trek from Earth to Mars (or elsewhere), but "life" from Earth may well have survived transportation to other worlds.

The prediction about specific alien forms having their origin on Earth is easily testable with genetic data, irrespective of whether those forms are humanoid. DNA sequence data like that shown in figure 3 can be used to construct evolutionary relationship diagrams such as that depicted in the top panel of figure 2, called "phylogenetic trees." As discussed earlier in the chapter, the forms with the most similar DNA sequences tend to have recent shared ancestors (e.g., dog-fox, marked with * in figure 2), whereas species with less similar DNA sequences tend to have shared ancestors further back in time (e.g., dog-seagull, marked with *** in figure 2). If, for example, Vulcans, Cardassians, and Denobulans are derived from recent human ancestors on Earth (e.g., from *Homo erectus*), and if we were able to acquire DNA sequences from those species, we should find that humans, Vulcans, Cardassians, and Denobulans are all more

Chapter 2

Human:	AGCGATTAGGGT
Vulcan:	AGCGATTATGGC
Cardassian:	AGCGATTAGGAC
Denobulan:	AACGATTAGGGC
Chimp:	AATTACTATGAT

Figure 4. Hypothetical DNA sequences from various alien forms, humans, and chimpanzees. Shading indicates identity with the human DNA sequence. Note that the four humanoids all have sequences more similar to each others' (differing at only one or two sites) than any do to chimpanzees'.

similar to each other in DNA sequence than *any* of those species are to chimpanzees or gorillas (see figure 4).

In addition to depicting evolutionary relationships, phylogenetic trees can be used to infer the *directionality* of movement. An identical approach is sometimes used in criminal cases for determining which person infected another with HIV.[31] If we assume transits are rare, either in the case of life to other planets or infection of HIV from one person to another, then a simple prediction is that the donor "population" should have some strains that are more similar in DNA sequence to the recipient population than some donor strains are to each other, whereas the opposite is not true (figure 5). Looking at the left panel of figure 5 as an example, all of Barb's HIV strain sequences have a recent shared ancestor, whereas John's HIV strain sequences have a more distant shared ancestor. Indeed, strain John04 has a more recent shared ancestor with the Barb sequences than it does with strains John01, John02, or John03. Hence, from this data we can be confident that Barb did not infect John; rather, John likely infected Barb. The same logic holds for the alien versus Earth species in the right panel of figure 5: all the alien forms have a recent shared ancestor, whereas the Earth forms have much more distant shared ancestors. The fictional phylogenetic tree depicted would suggest perhaps a single exodus of human ancestors from Earth, more recent than the split of human and Neanderthal, led to the alien forms depicted. These examples show just some of the many uses of phylogenetic approaches applied to DNA sequences.

The hypothesis that ancient life from Earth had gone elsewhere was not limited to *TOS* "The Paradise Syndrome" within *Star Trek*. In *VOY* "Distant Origin", the reptilian alien Voth species were inferred to have evolved from dinosaurs on Earth and thus be related to humans and other life on Earth. The episode suggested that the Voth may have descended from hadrosaur dinosaurs on Earth, and while a few real-world studies have suggested (and as Captain Janeway also speculates in the episode) a few hadrosaur populations may have survived slightly past

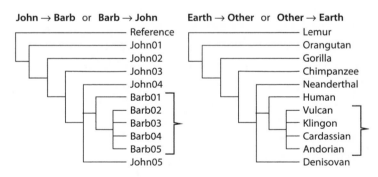

Figure 5. *Left panel*: phylogenetic tree derived from five individual HIV strain sequences each acquired from hypothetical individuals John and Barb. The bracket highlights that Barb's HIV strain sequences are all more similar to each other than to any of John's. *Right panel*: phylogenetic tree from DNA sequences acquired from various Earth and alien species. The bracket highlights that the alien species are all more similar in DNA sequence to each other and to humans than humans or aliens are to other Earth primates.

Earth's mass dinosaur extinction,[32] many other studies question their approaches and conclusions.[33] This episode erred on a few paleontological details, such as the computer labeling a holographic recreation of an ancient amphibian as being of genus *Eryops* (it appears more similar to a gorgonopsid) and the genus being assigned to the wrong geologic period (Devonian rather than Permian). However, more importantly, the episode's use of genetic markers to demonstrate shared ancestry between the Voth, humans, and reptiles from Earth was appropriate. *Voyager*'s holographic doctor likely employed an approach similar to that described above, assuming his program was given a rigorous background in evolutionary biology.

The essential element not discussed at all above is how life from Earth could get fully established on another world within tens or hundreds of millions of years. Not only would a life form have to arrive safely on another world, but an ecosystem on the new world would have to support this life. What would the source of nutrients and energy be for an ancient humanoid or reptile? In *TOS* "The Paradise Syndrome," it appeared as though the Preservers transported not only humans but also plants (pine trees, orange blossom, honeysuckle) and other animals. Establishment of such an ecosystem on an alien world would also be quite challenging, as the process of "terraforming" would certainly be a major undertaking. As my focus is on evolutionary relationships in this chapter, I do not delve further into exploring whether or how this might be possible, but I merely highlight that evolutionary feasibility does not equate to technical feasibility.

Nonetheless, if life arises from nonlife very infrequently, the first extraterrestrial life we encounter may well be life that originated on Earth. This observation is reminiscent of the last line of Ray Bradbury's *The Martian Chronicles*, when humans who had moved to Mars sought "Martians," and observed them as their own reflections in the rippling water of a Martian canal.***

CLOSING REMARKS

This chapter has presented a general introduction to evidence for evolution, misconceptions about evolution, and testing various hypotheses regarding the origin of life on Earth and elsewhere using genetic data and evolutionary principles. While the writers of *Star Trek* occasionally slipped with some incorrect details or by reinforcing some misconceptions about evolution, they deserve praise for wholly embracing the knowledge that modern species (those on Earth and those elsewhere) share ancestors, and that evolution occurs both through changes within lineages and through formation of new species/lineages. Frankly, they do a better job of embracing evolution than biology courses in several high schools in the United States.

In considering the evidence for evolution and testing the hypotheses above, this chapter has leveraged DNA sequences and principles of inheritance. The next chapter delves into DNA, mutations, and inheritance in more depth, both in the science and as discussed in *Star Trek*. I will also return to some specific *Star Trek* episodes already mentioned (e.g., *TNG* "Genesis"), but explore the proposed genetic mechanisms and their feasibility in more detail.

***   Bradbury R. *The Martian Chronicles*. Doubleday, 1950.

CHAPTER 3

DNA

Evolution's Captain

All offspring inherit traits from their parent(s). This inheritance exists both in the specific sense that offspring tend to look more like their parents than other random individuals, but also more generally in the sense that species breed true; for example, cows do not give birth to sunflowers. The same information molecule, DNA, confers similarity in traits between parents and offspring as well as species identity. Importantly, variation in this information constitutes the raw material for evolution, as I will discuss at greater length in the next chapter.

Star Trek mentions DNA, genes, mutations, and other facets of inheritance throughout the series that aired after *TOS*. Major advances in genetics came about during and after the earlier series aired on television (e.g., new genetic engineering technologies), some of which may make depictions in the series that once seemed unfeasible now possible. In the next few sections, I will cover briefly some basics about DNA, how a cell's DNA is translated into traits, how the DNA is passed on to new cells, and how DNA changes over time both within individuals and between generations.

DNA: WHAT IS IT?

VOY, Season 1, Episode 10, "State of Flux"

Crewmember Seska (who appears to be a Bajoran species member) has been injured, and upon examining her blood, the ship's nurse notices that her blood is lacking all the common Bajoran blood factors. The holographic doctor argues that Seska had been "genetically altered" and is likely instead a Cardassian species member. When confronted, Seska insists she is Bajoran but had contracted a viral disease as a child. She says that she received a bone marrow transplant from a Cardassian, and that transplant makes her appear to be Cardassian. The doctor maintains that the genetic markers in her blood clearly indicate she is Cardassian, and no virus or bone marrow transplant would explain them away.

DNA, or deoxyribonucleic acid, is the molecule within cells that encodes much of an organism's structure and behavior. The information in DNA is stored as a code, with each unit being one of four possible nucleotide* blocks, abbreviated A, C, G, and T. The identity and order of a set of nucleotide blocks is called a "sequence"—for example, "ATGAAC" would be a short sequence. DNA is packaged in the cell as one or more long, threadlike structures called "chromosomes," and the full set of DNA instructions in a cell is called its "genome." The number of nucleotides in the genome can be quite large in some species (two copies of approximately 3.1 billion units in humans). Organisms made up of many

---

* The word "nucleotide" is used several times in the various *Star Trek* series, beginning with *TOS* "Let That Be Your Last Battlefield."

cells—like plants, animals, and fungi—typically have nearly identical genomes in each cell of an individual. With modern technologies, we can now study partial or whole genome DNA sequences from an individual with vanishingly small amounts of their tissue or bodily fluid (e.g., saliva).

Researchers obtain DNA sequences for a variety of purposes, such as determining if a person is another's true parent, identifying an individual's country or region of descent (or, for nonhumans, their breed or species), testing whether a person was present at a particular location via cells that person left behind, and many more. As discussed in chapter 2, similarity in DNA sequence is hierarchical with relationship: individuals' DNA sequences are most similar among their own cells, more similar to those of their immediate families than to nonrelatives, more similar to other humans than to other primate species, more similar to primates than to other animal species, etc. Databases with thousands of whole- or partial-genome DNA sequences exist online, many of which are easily searchable.[1] As a consequence, one can match a test DNA sequence on a personal computer to sequences from online databases to infer a test subject's identity, heritage, or species.[2]

The various *Star Trek* series aired during a period when scientists' understanding of and ability to work with DNA were steadily expanding, so a progression occurs across time in the fraction of episodes with references to genetic terms (see figure 6). *TOS* aired in the late 1960s, a few years before scientists even knew how to obtain DNA sequences from individuals.[3] *TOS* episodes generally do not mention DNA even in cases where it would have made sense for the plot.[†] For example, Doctor McCoy concluded

†   The only single mention of "DNA" in *TOS* was in the episode "The Immunity Syndrome."

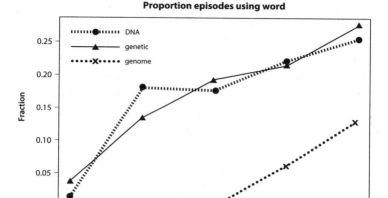

Figure 6. Proportion of episodes of each *Star Trek* series mentioning the word "DNA," "genetic," or "genome." Years indicate the last year the series aired.

teenager Charlie was human based on "the development of his fingers and toes" rather than based on a DNA test,‡ and Captain Kirk used a computerized voice test rather than a DNA test to determine if actor Anton Karidian was actually the fugitive Governor Kodos.§ Similarly, the project to obtain a complete human genome DNA sequence began in earnest in the 1990s, with the first nearly complete genome sequence published in 2001.[4] Correspondingly, the word "genome" appears several times in *VOY* and *ENT* (airing 1995–2001 and 2001–5, respectively) but not in the older series *TOS*, *TNG*, or *DS9* (figure 6).**

‡ *TOS* "Charlie X."
§ *TOS* "The Conscience of the King."
** It is used once in *TNG*, but as a proper noun referring to "Genome Colony."

By incorporating newer genetic approaches, the later *Star Trek* series depicted DNA testing successfully many times. For instance, Doctor Mora in *DS9* "The Alternate" had two DNA samples from unknown life forms and notes regarding them, "It's not the same entity. The nucleotide sequences are entirely different.... At best they could be distant cousins." This observation is entirely reasonable; for example, personal genotyping companies today store DNA samples and help customers identify other customers that are likely distant relatives—say, third cousins. The discussion in the *VOY* episode referenced at the beginning of this section regarding the Cardassian spy Seska was also reasonable. If she is indeed a Cardassian rather than a Bajoran, a DNA test would readily identify her as such, presuming that DNA sequences are available in the database from both species for comparison. Her argument about the bone marrow transplant was also reasonable—blood cells are made from bone marrow, so if Seska had obtained a bone marrow transplant from a Cardassian, it makes sense that at least some of her (white) blood cells would have Cardassian DNA. Perhaps the doctor was then able to reject her explanation because there was a complete absence of Bajoran blood factors, rather than a mix. Less clear, however, is what the doctor meant regarding Seska being "genetically" altered to appear Bajoran. This comment may refer to manipulations that affected which genes were turned on or off—the topic of the next section.

DNA: HOW DOES IT WORK?

Star Trek: Nemesis (2002) movie

Shinzon is a human who looks strikingly similar to Captain Picard of the *USS Enterprise-D*. Upon examining a blood sample, the crew discover that he is a clone of Captain Picard. Doctor Crusher notes, "Shinzon was created with temporal RNA sequencing. He was designed so that at a certain point, his aging process could be accelerated to reach [Picard's] age more quickly. . . . But when the temporal sequencing wasn't activated his cellular structure started breaking down. He's dying. . . . Nothing [can be done for him] except a complete transfusion from the only donor with compatible DNA."

Scientists definitively showed in the 1940s and 1950s that DNA carried hereditary information,[5] but proteins, made up of long strings of amino acids, are the workhorses of cells. Some proteins provide structure and support for cells, some transport molecules or signals within and between cells, some—enzymes—catalyze chemical reactions in cells, and some—antibodies—neutralize pathogens. Hence, much of a cell's type (e.g., skin versus nerve) or an organism's identity (human versus chimpanzee) is determined by which proteins are present and in what quantities, the specific amino acids making up these proteins, the timing of protein production, etc.

One of the most fundamental principles in molecular biology, termed the "central dogma of molecular biology," is that information associated with inheritance flows largely in one direction between three chemical domains: from DNA to RNA to proteins.[6] Segments of the DNA-based genome (what we typically call "genes") are used in cells as templates for production of strands of the nucleic acid RNA, also made up of nucleotide blocks. Different types of RNA serve different functions, but the nucleotide identity and order in one type of RNA determine which amino acids are assembled to make particular proteins. The central dogma highlights the typical directionality of the unit-by-unit transfer of information from DNA, used as hereditary information transfer, to RNA, some kinds of which act as intermediates, to protein, used in cell structures and functions. While some movement of information from RNA to DNA is permitted, the original information cannot go back to RNA or DNA from the amino acids in proteins.

Given the DNA code is largely identical across all our cells, why do some cells in an individual develop into muscles, some become skin, some become nerve tissue, etc., each with different physical features? Part of the answer is that different genes make RNA at different times and in different cells. For example, some genes turn on in response to the external environment (e.g., to "heat"), some respond to internal cues (e.g., presence of hormones or other compounds), and some turn on or off with age or season or even time of day.[7] As a result, even though the same DNA code is present in all cells, different genes are "activated" across cells in different degrees and/or at different times, resulting in different protein abundances and overall different cell forms in the end.

I will use the analogy of reading a book for much of this chapter, but I ask the reader to remember that analogies can be taken only so far. One can think of genes as chapters of a book, with the whole book being the full genome's DNA code in cells. One can think of reading aloud as equivalent to making RNA, and the eventual proteins are like the products of what is read aloud. Imagine that some chapters are only read aloud in cells exposed to ultraviolet light, while other chapters are read aloud in cells not exposed to such light. Hence, even though the "book" is the same, different proteins are produced in different cells because of differences in which genes are activated to make RNA. Further, some genes are fine-tuned to produce more or less RNA (rather than just turn on or off) in specific cells at specific times, similar to a chapter read more or less loudly in these different cells.

Sometimes, a single gene can even make different proteins. Some genes have segments of DNA called "introns" inside them that do not encode amino acids—these stretches of instructions literally get "cut out" of the RNA after it is produced and therefore do not affect which amino acids get assembled. One can conceive introns as "footnotes" in the book of the genome that are not always read aloud. However, one way a gene can make different proteins is if a segment is sometimes cut out and sometimes not (varying among cells or individuals). This variation in whether introns are cut out is quite common in humans; one study estimated that 90%–95% of genes bearing introns exhibit such variation, causing them to make different amino acid sequences, resulting in different proteins.[8]

The *Star Trek* series understandably do not delve deeply into molecular biology or genome structure. However, the terms explained here do crop up. The quote at the beginning of this section refers to Shinzon's aging being accelerated through some modification of RNA, presumably referring to how much RNA is produced from specific genes—how loudly those chapters are read—rather than changes in the RNA's nucleotide sequence. Again, specific genes' RNA production changes with advancing age, and these changes can be manipulated; for example, restricting the number of calories one consumes reduces age-related changes in RNA production.[9] Such manipulation could be done with targeted drugs as well, and if such a manipulation were done poorly or incompletely, the procedure could result in Shinzon having severe health issues. I am uncertain why a transfusion from someone with identical DNA—Picard—would help his situation, except perhaps through differences in epigenetic marks (discussed later in this chapter).

Many other *Star Trek* episodes mention gene activity. For example, Doctor Crusher mentions a "dormant gene" in *TNG*

"Genesis,"[††] and the doctor mentions that one of Commander Chakotay's DNA segments has been "hyperstimulated" in *VOY* "Scientific Method," presumably illustrating decreased and increased RNA production respectively. The *TNG* episode "Genesis" mentions introns, too. Lieutenant Commander Data describes introns as "evolutionary holdovers, sequences of DNA that provided key behavioral and physical characteristics millions of years ago, but are no longer necessary." Data seems to be incorrectly equating introns with vestigial genes (discussed in chapter 2). Because introns are often cut out of RNA before proteins are produced, they are likely to accumulate mutations over evolutionary time since natural selection would not eliminate such mutations (discussed in the next chapter). Thus, the information content in introns would be more likely to change over time than to retain information about earlier evolutionary states. I will return to the subject of mutations in the final section of this chapter.

DNA: HOW IS IT TRANSMITTED AND MANIFESTED?

VOY, Season 3, Episode 20, "Favorite Son"

Ensign Harry Kim of the *USS Voyager* begins to exhibit unusual behaviors and develops spots on his face. The Taresians (an alien species) have tricked Ensign Kim into believing he is a member of their species, claiming that other Taresians implanted his human mother with him as an embryo. They insist that the reason for his changes is that his Taresian genes are becoming active. The holographic

††   Though this reference is odd in that she says it is a "T-cell in your DNA," which is, at best, unclear.

doctor identifies that Ensign Kim indeed has Taresian DNA fragments, which had been missed in past scans, noting, "These fragments have apparently been disguising themselves as the recessive elements that normally exist within all DNA, but over the past few days they've become increasingly dominant over his human genes." Later, the doctor finds that scans of Ensign Kim from several weeks earlier do not have this Taresian DNA. Apparently, a retrovirus (engineered by the Taresians) infected Ensign Kim without his knowledge weeks earlier, and it transferred Taresian DNA into his cells.

Two types of DNA transmission occur within multicellular organisms, like virtually all plants, animals, and fungi. First, we continually make new cells within our bodies. When a cell divides, the DNA code from the original is passed along into the new cells, largely unchanged (barring new mutations, discussed later in this chapter). Second, we combine elements of our genome with others' to produce offspring. Most animals have two copies of every gene in their genome, one acquired from their mother and one from their father. These two copies are usually similar in nucleotide sequence but not always identical. For instance, within a gene affecting eye color, a mother may have nucleotide variants associated with making brown eyes, whereas a father may have variants associated with blue eyes. One of the two copies of each gene from the parent's genome goes into each of the gametes (egg or sperm for humans, seed or pollen for plants, etc.). When fertilization occurs, uniting the one copy from the mother and one from the father, each offspring then receives two copies of all genes again. These offspring then pass on either their mother's gene copy or their father's (but not both together) in each of their gametes.

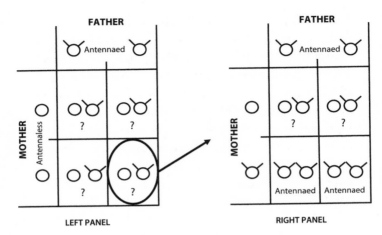

Figure 7. Fictional Punnett squares depicting gene variant combinations in offspring from antennaed father and antennaless mother (*left panel*) and from one of their female offspring with an antennaed father (*right panel*).

The inheritance of variants at specific genes is often depicted using "Punnett squares," as shown in figure 7 with a fictional gene affecting whether individuals have antennae. In this case, the left panel depicts potential offspring from a father possessing two gene copies for "making antennae," and a mother possessing two gene copies for "not making antennae." The father's two copies are depicted on the top, and the mother's two are depicted on the left. The four boxed areas show all the possible combinations of the mother's and father's gametes that might come together in offspring, which in this case are all identical (one gene variant for making antennae, one gene variant for being antennaless). Let us then take the offspring in the lower right box on the left side, and have her produce offspring with another male possessing two copies for having antennae (right panel). Now, the four boxed areas show two different possible outcomes: half the offspring will have one variant for producing antennae and one variant for being antennaless; the other half will have two copies of the variant for producing antennae.

Punnett squares depict inheritance but they do not say anything about the physical manifestation. Presumably, a man with two antenna-producing gene copies will have antennae, and a woman with two "antennaless" copies will not. But what about the offspring in the left panel, with one antenna-producing and one antennaless variant? The manifestation is not obvious. In some cases, the offspring may be intermediate—for example, having miniature antennae. In other cases, one variant dominates the other, such that the offspring in the left panel may all have full-length antennae like their father, for example. In that example, we say that the antenna-producing variant is "dominant," since having just one copy causes the same physical manifestation as having two. The antennaless variant in that case is called "recessive." Importantly, this does *not* mean that the antenna-producing variant is healthier or better or more common in the population; it just means that having one copy causes the same physical manifestation as having two.

What is the cause of such dominance? As discussed in the preceding section, proteins are the ultimate products of genes, and they are what cause the physical manifestations. The antennaed and antennaless variants may differ in the amino acids, and thus proteins, they produce. Alternatively, they may result in different amounts of basically the same protein being produced. In the latter case, let us imagine that producing 1,000 units of RNA (or more) results in enough protein being made to make an individual grow antennae. Let us imagine that the antenna-producing gene variant produces 900 units of RNA, and an "antennaless" gene variant produces 200 units. An individual with two antenna-producing variants will produce 1,800 units of RNA, well above

the 1,000-unit threshold, and grow antennae. An individual having two antennaless variants will produce 400 units of RNA,
well below the 1,000-unit threshold, and grow up without antennae. If an individual has one of each variant, that individual
will produce 1,100 units, above the 1,000-unit threshold, and still
grow antennae. Going back to the book-reading analogy, one can
think of both copies of the book being read simultaneously, with
the dominant copy of that gene/chapter being read so loudly
that a listener does not notice what was said by the recessive copy.

Dominant and recessive variants are accurately described in
Star Trek. Klingon forehead ridges are described as dominant
to smooth human foreheads in *VOY* "Lineage," which is consistent with Klingon-human hybrid B'Elanna Torres having such
ridges and being able to transmit this trait to her daughter.[‡‡]
In contrast, Skagaran forehead ridges are described as recessive
to smooth human foreheads in *ENT* "North Star," which explains
why a character with one Skagaran grandparent and three human grandparents could not have these ridges—she cannot inherit two gene copies for ridges if there is only one copy among
her parents. At most, she would have one copy for ridges, which
would be masked by the human unridged variant. Borg drones
who could access a virtual setting called Unimatrix Zero are said
to have a "recessive mutation."[§§] Given the mutation is recessive,
a drone presumably needs to have two copies of the mutation,
rather than just one, in order to access this setting, which may
help explain why few Borg are able to.[***]

[‡‡]   Another human-Klingon hybrid is K'Ehleyr from *TNG*, who also has forehead ridges.
[§§]   *VOY* "Unimatrix Zero."
[***]   Borg are made up of other species that have been assimilated, so how
would so many species share the same rare recessive mutation? One possibility

Returning to the *VOY* example at the beginning of this section, one remarkable aspect is that the doctor refers to Ensign Kim's Taresian DNA "becoming increasingly dominant." A change in dominance may seem odd, but researchers have documented examples of environmental effects altering the dominance of particular gene variants.[10] However, the usage here appears different: the doctor seems to imply that the genes' becoming "dominant" has affected the detection of the DNA itself (hence why the Taresian DNA was not detected in earlier scans) rather than merely its physical manifestation. This is a common misconception. If one performs DNA sequencing (a logical "scan"), one will certainly detect both gene copies irrespective of whether they are dominant or recessive. Returning to the book analogy, regardless of how loudly words in a book are read, the words are still present on the page. Presumably the doctor used DNA sequencing later in the episode to note the absence of Taresian DNA in Ensign Kim's earlier scans.[†††] Perhaps the first set of "scans" mentioned focused on RNA or proteins, and RNA production may have changed over time or in response to an external stimulus. Changes in DNA sequence (including via retroviruses) and in RNA production are the focus of the next section.

---

is that the mutation is a byproduct of the Borg assimilation process rather than something extant within the species. *TNG* "The Best of Both Worlds, Part 2" does mention DNA being "rewritten" as part of assimilation, so this explanation is consistent with the *Star Trek* canon. Perhaps the rewriting occasionally introduces not one but two errors, thereby resulting in access to Unimatrix Zero.

[†††] In *ENT* "Cold Front," set long before the whole *VOY* series, Dr. Phlox notes that their standard imaging device "allows us to examine your genetic structure," and T'Pol is able to scan the DNA of aliens in another spaceship in a few seconds in the next episode, *ENT* "Silent Enemy."

DNA: HOW DOES IT CHANGE?

VOY, Season 2, Episode 15, "Threshold"

Lieutenant Tom Paris successfully breaks the record for fastest flight ("warp 10").[‡‡‡] Upon returning, he becomes gravely ill. The doctor observes that he has a high rate of cellular mutation. He also undergoes many physiological changes, and he even grows a second heart as well as other new organs. Talking with the captain, the doctor notes, "His body is going through some sort of mutation. His DNA is rewriting itself." Paris becomes psychologically unstable and escapes, kidnapping the captain, and forcing her to go with him at warp 10. Following this flight, the two crew members both change into entities resembling giant salamanders within a few days.

Given the common ancestry of all life on Earth, we must infer that genomes have changed over evolutionary time to produce the physical variation within and between species alive today. Multiple kinds of genomic changes can produce such variation. Random changes in a DNA nucleotide sequence—"mutations"—can happen during cell division within an individual or during the formation of gametes, in each case resulting from accidents during cell division or following exposure to chemicals or radiation. One can think of these mutations as "copy errors." Perhaps the best-known general category is a mutation in the DNA sequence at a particular nucleotide site. Hence, if a segment of a gene's DNA sequence was AATAGC in the parent cell, it might

[‡‡‡]   Speeds above warp 10 were achieved in *TOS,* so presumably there was a redefinition of warp speeds.

be AATGGC in the new cell, resulting from a mutation in the fourth nucleotide (from A to G). A second general category is "structural" changes. If we again think of the genome's DNA sequence like a book, then structural changes could include deleting a sentence, duplicating an existing sentence, reorganizing sentences, etc. Fundamentally, these changes are all considered "mutations," and changes resulting from either of these categories can be passed on to subsequent generations. Unsurprisingly, the more times one makes copies from earlier copies, the more likely one is to produce mutations, which potentially lead to cells becoming cancerous[11]—one reason cancers may increase greatly in frequency with age.

Whether one perceives mutations as happening "frequently" or "rarely" depends on the scale of examination. Within an individual human, mutations occur as cells divide and are then passed on to new cells. While the exact rate is difficult to measure (although recent development of single-cell DNA sequencing approaches have helped dramatically),[12] we now know that depicting individuals as having exactly the same DNA code in all of their cells is an oversimplification. Some cancers in particular— for example, melanomas—are associated with especially high rates of mutation as cells divide within a body.[13] Many species have mechanisms to repair some types of mutations that arise or to force individual cells with mutations to die rather than make more mutant copies, but these mechanisms remain imperfect, and mutations do propagate. In terms of a "between-generation" mutation rate (i.e., changes between parent and offspring, likely resulting from mutations in the parents' gametes), single nucleotides change at a rate of 1.1×10^{-8} per generation, or roughly one change per 100 million sites per generation.[14] On the one hand, this figure seems like an incredibly low rate, but on the other, given there are more than three billion sites in the human genome, that means each person acquires on average roughly 70 new mutations (35 from each parent), and some studies suggest this figure may even be an underestimate.[15] Hence, we are all "mutants" relative to our parents' genetic makeup.

Occasionally, individuals even acquire DNA from viruses. For example, retroviruses are viruses that insert RNA and enzymes that produce DNA into host cells, and the new DNA can become integrated into the host's genome. HIV is an example of a retrovirus; its RNA inserts into human cells and it changes their genome's DNA sequence, thus driving these human cells to produce more of the virus. I discuss some cases of getting DNA from other kinds of species in the Appendix.

A final category of change I will mention briefly falls under the loosely and inconsistently defined umbrella of "epigenetics."[16] As mentioned earlier in this chapter, different genes may be turned on or off in particular types of cells or at specific times. One mechanism for this variation is that chemical marks—one example being a "methyl group"—sometimes bind to specific sections of DNA, preventing RNA from being produced there. To go back to our book analogy, one can think of this as striking out a particular passage so it is not read in some cells. The words are still present, but one does not read them. Many internal and external factors affect whether such marks are placed on DNA in particular cells; for example, exercise affects their placement on muscle cells,[17] aging changes these marks,[18] and drugs can affect them.[19] Importantly, these marks may persist when the cell divides, so they are sometimes inherited among cells within an individual's body. Such marks were typically thought to be lost when gametes are formed in mammals and therefore to not be passed on to offspring in the way genetic mutations are. While some such marks were known to be passed on to offspring in plants, a few studies in mice have now shown that some epigenetic marks can be passed on to offspring,[20] and these observations have attracted great interest.[21] Other hints exist for such non-DNA-sequence-based transmission in mammals, such as female rat offspring having diabetes-like symptoms if their fathers were fed high-fat diets.[22] Scientists have not yet come to consensus on whether these cases are rare, unimportant exceptions (e.g., even in several of the cases known, transmission disappears within a few generations, unlike DNA-sequence-based mutations)[23] or whether we will eventually find many traits inherited in this manner over multiple generations.

Coming back to *Star Trek*, various types of mutations or genetic changes are mentioned, but the specific molecular nature is not usually discussed. For instance, Doctor Bashir speculated that Jem'Hadar warrior Goran'Agar had a mutation that prevented him from becoming addicted to the drug ketracel white in *DS9* "Hippocratic Oath." Retroviruses inserting DNA into genomes were mentioned, such as the viruses affecting a group of Klingons in *VOY* "Prophesy" and the one causing Ensign Kim to think he was a Taresian in *VOY* "Favorite Son" mentioned earlier in this chapter. Several episodes of *VOY* (e.g., "Phage," "Faces") explored the plight of the Vidiians, a species that suffered for thousands of years due to a "phage" disrupting their genetic code and cell structures. The term's usage in the episodes appears somewhat different from the typical scientific use today (short for "bacteriophage"), which refers to a virus that attacks and replicates within bacteria. Instead, it may refer to a retrovirus, which presumably is specific to Vidiians since there was no mention of it affecting other species.

One common mistake in science fiction is assuming that mutations arise in an individual in many cells almost simultaneously and cause nearly immediate manifestation throughout the body. An adult human body has trillions of cells, so getting the same mutation in each cell would be an astounding feat. Still, several *Star Trek* episodes depict virtually complete genetic and physical transmutations of characters into a new species within a few days following parasite infection or other encounters (e.g., *TNG* "Identity Crisis," *VOY* "Vis à Vis," *ENT* "Extinction"). An extreme example was depicted in the *VOY* "Threshold" example highlighted at the beginning of this section. Rather than developing random defects or cancers as one might expect, Lieutenant Paris and Captain Janeway metamorphose into amphibians in a few days. First, genetic mutations should not be so repeatable—even if induced in the same way, mutations are random changes in the DNA code. Returning to the book analogy, one can induce a higher rate of change by closing one's eyes and randomly taking a pen to a page to alter letters, but taking a pen to a page a second time the same way will be unlikely to induce the same word change. Lieutenant Paris and Captain Janeway should not have had precisely the same genetic changes or ultimate manifestation.§§§ Second, the speed of these changes, happening within hours or days, is remarkable. Even if someone instantly had an altered genetic code in all their cells in their current body, how long might it take for the cells to form a second heart? In human adults, less than 1% of heart muscle cells turn over each year.[24] One could not simply grow another heart in a few hours or days, even if the specific mutations arose that might encode such a process, much less become an amphibian.

§§§ The episode dialogue suggests that they may have changed into a future form of human evolution, perhaps since the warp 10 flights somehow had them transcend linear time. If it is truly a future form, and not an "expectation" of the future (see chapter 4 on issues with predicting evolution), then that may partially explain the similarity and the goal-oriented nature of the changes. However, this hypothesis would not explain why these evolutionary changes would happen to an individual in the first place: when we go to a distinct place or time, we do not suddenly change form to match other entities there.

However, changes in how much RNA a gene produces (perhaps via epigenetic effects) may be (slightly) more likely than mutations to create some of the manifestations depicted in *VOY* "Threshold."[25] An external stimulus may cause many cells in an individual to alter their RNA production simultaneously and in the same manner. In some cases, one need not even wait for cell division for manifestation to occur—the cell that is already there may immediately start producing different proteins. For instance, Doctor Pulaski and others are exposed to an antibody in *TNG* "Unnatural Selection" that alters their DNA in a way to promote rapid aging. Although the premise is a stretch either way, it seems more likely that such antibodies would act as an external stimulus affecting RNA production in cells rather than somehow directly changing the DNA sequence of multiple cells in multiple people in precisely the same manner.****

DNA: RECOMBINATION

DIS, Season 1, Episode 1, "The Vulcan Hello"

The *USS Shenzhou* approaches a star system and seeks to investigate a curious area that is unapproachable by spaceship. Commander Michael Burnham volunteers to fly in a space suit with a thruster pack to investigate. Science officer Saru notes that the radiation coming from the stars may cause her DNA to unravel after 20 minutes. Commander Burnham nonetheless remains away for longer

---

**** This interpretation makes more scientific sense, but the episode explicitly shows Lt. Cdr. Data pointing to a computer screen with DNA sequences and suggesting that two nucleotides were transposed as a result of the antibody. Similar epigenetic effects may explain the crew's radiation-induced aging in *TOS* "The Deadly Years" and its subsequent cure with the hormone adrenaline.

than 20 minutes. Three hours later, she is lying in a device having an "antiproton regimen" applied. Commander
Burnham bursts out of the device to see the captain, and
the medical officer chases her, exclaiming, "The recombination process is nowhere near finished! I need you back
in the antiproton chamber. Do you understand the effect
of genetic unspooling? You don't want to die that way."

This momentary scene from *DIS* has a lot of genetics and molecular biology to unpack, so I will focus this whole section on
it. First, as mentioned in the preceding section, radiation causes
DNA damage. Such damage may include mutations that change
the DNA sequence, such as those described earlier, but it may
also involve literal "breaks" in the DNA that can be problematic
to a cell, potentially leading to cell death.[26] Fortunately, our cells
have safeguard mechanisms to deal with such damage, including "recombination."

Geneticists use the term "recombination" in two different, albeit related, ways.[27] High-school and college biology classes refer to recombination associated with shuffling variants at genes
while gametes are formed. For example, imagine two genes called
A and B. Imagine I inherit variants A1 and B1 at these genes from
my father's sperm, and I inherit A2 and B2 from my mother's egg.
If one of my gametes has the combination of variants A2 and B1
(thereby having one variant inherited from my mother and one
from my father), then a "recombination" event has occurred when
my gamete was formed, and I am passing on a different combination of gene variants than either of the combinations I inherited from my parents. Typically, genes closer to each other along a
chromosome recombine less frequently than genes far away. Such
recombination generates genetic diversity—more combinations
of gene variants in the population—which in turn generates a lot

of the diversity that we see in how members of a species look and act. However, while this process is important for evolution (see the Why Sex? section of chapter 5), this is not the form of recombination mentioned in the episode, since the threat appears to be to Commander Burnham's health directly and not to something regarding gametes or offspring.

The other form of recombination, sometimes called "mitotic recombination," happens within the body's dividing cells. This process repairs DNA damaged by various agents including ultraviolet radiation, gamma rays, X-rays, and other forms of radiation. It can involve repairing the broken DNA by copying the DNA sequence from an undamaged chromosome as a template for repair, hence also creating a new "combination." This form of recombination is much more likely to be what the treatment is facilitating in Commander Burnham in the medical bay.

One might interpret the machine in the medical bay as using "antiprotons" in some way to aid Commander Burnham's cells' engagement in DNA repair through mitotic recombination. Antiprotons are found in cosmic rays and have been evaluated as a component of cancer therapy.[28] Perhaps they were being used in this episode in a targeted way to stimulate production, repair, or other actions of some of the proteins associated with mitotic recombination (see below).

How might "spooling" be involved? Unlike simplified thread-like depictions in some textbooks, the two meters of DNA in every cell is tightly is "wound up" around proteins called histones. Hence, much of the genome is indeed "spooled" in its normal state, as implied in the quote. However, for the cell to repair DNA breaks, the histones surrounding the DNA break must first be removed to allow repair machinery access to the damaged DNA.[29] The radiation dose received by Commander Burnham likely resulted in thousands of DNA breaks, and significant "unspooling" of the DNA around those breaks was thus required for repair. If the DNA remained in the unrepaired and unspooled state, it would indeed have catastrophic consequences for the commander's cells and overall health. Additionally, histones are chemically modified during the DNA damage response process in ways that foster repair.[30] Conceivably, radiation could damage the histones themselves, leading to inadequate DNA repair responses as well as perhaps some failures in proper spooling.

Overall, I might infer that Commander Burnham's exposure to radiation resulted in many DNA breaks, and her cells then began to try to repair those breaks, thereby requiring unspooling much of the DNA. However, the damage to her DNA, and perhaps also to her histone proteins, was so extreme that a medical intervention was required to complete the repair and get the DNA back into its normal spooled state. While I am arguably overly generous to the writers with the description above of how elements in the quote may reflect real biological processes, the example nonetheless provides a great entry point for introducing a bit more cellular and molecular biology to reader.

GENETIC ENGINEERING: DNA'S FINAL FRONTIER

This chapter has focused on exploring how genetics "works" in natural systems. Before closing, I turn briefly to how genetics is manipulated for various purposes, both in the real world and in *Star Trek*.

Scientists, science fiction writers, and the public are all very interested in—and many afraid of—the promise and prospects of genetic engineering. Simply put, genetic engineering involves manipulating an organism's genome directly with some sort of technology. Like natural mutations or transfers, genetic engineering could involve changing a nucleotide sequence, removing a gene, adding a gene, or rearranging the placement of genes in a genome. Many reasons exist for genetic manipulation: to repair a disease-causing mutation; to make crops less susceptible to pests, pathogens, or environmental stress; to conduct research on the function of genes; etc. However, science fiction—and occasionally public dialogue—often falls back to more controversial uses, such as making "designer babies" (sometimes with unexpected effects), supersoldiers, clones, or biological weapons.

Star Trek explored each of these themes over the various series, though referring to genetic engineering as DNA "resequencing" (see below). Doctor Julian Bashir in *DS9* is himself a product of genetic enhancement in intelligence, stamina, vision, and reflexes.†††† His case seemed successful in all areas, while others who were genetically manipulated developed unanticipated personality quirks (*DS9* "Statistical Probabilities") or became violent (*TOS* "Space Seed"; *ENT* "Borderland," "Cold Station 12," "The Augments"). The Jem'Hadar that appeared in many episodes of *DS9* were genetically engineered supersoldiers produced by the Dominion (an aggressive interstellar state). While not exactly genetically engineered since no deliberate genetic alterations were made, a genetic copy ("clone": see chapter 5) of the legendary Klingon warrior Kahless was nonetheless made many years after his death using the original's genetic samples in *TNG* "Rightful Heir." Various species have developed genetically engineered pathogens as biological weapons; for example, *TNG* "Chain of Command,"†††† *DS9* "The Quickening," *VOY* "Child's Play."

---

†††† *DS9* "Doctor Bashir, I Presume."
†††† Though the example here turned out to be false within the episode.

While genetic engineering has been around more than 40 years already, the past decade heralded major advances in artificial "transgenic" technologies, wherein scientists can now quickly and reliably target and delete or replace particular genes with only slightly more difficulty than depicted in science fiction. Several means for transgenics are commonplace in molecular biology laboratories, but most recent attention has focused on the CRISPR/Cas9 system due to its simplicity, ability to work in many species, and low cost. Using a mechanism similar to one bacteria and archaea use for defense against viruses, researchers can combine a few simple elements—an enzyme that cuts DNA, and RNAs that target where the enzyme should cut—to precisely excise a section of a gene or replace a stretch of DNA with another.[31] The flexibility of this system makes it easy for scientists to "edit" genes, and application to human embryos has begun,[32] demonstrating the feasibility of correcting disease-causing genetic defects. Obviously, application of such approaches to humans comes with a litany of ethical and legal questions, and our community (and our fiction) will doubtless struggle with this for years to come. Indeed, an era in which "DNA resequencing" becomes common may already have begun. Fortunately, we did not—and I am confident we will not—genetically engineer supermen (Khan being one example)[§§§§] that cause millions of deaths, as predicted to have occurred in the 1990s in *TOS* "Space Seed."

§§§§   *TOS* "Space Seed," *Star Trek II: The Wrath of Khan* (1982), and *Star Trek Into Darkness* (2013).

CHAPTER 4

CHANGE OVER TIME
Drivers of Evolution

Despite descending from a shared ancestor, life forms on Earth today vary greatly in appearance and behavior. In the preceding chapter, I explained that much of the physical variation within and among species is encoded in the DNA of individual organisms, and that mutations are what ultimately generated the variation in DNA that exists across species on our planet. Processes that cause new mutations to spread or to be eliminated from a species are the drivers of evolution and the subject of this chapter.

Terms like "evolution" and "natural selection" have become common parlance in science fiction, but their usage is frequently incorrect, imprecise, or misleading. In this chapter, I will first elaborate what does and does not constitute evolution, noting some misuse of the term. Then, I will discuss examples of a few evolutionary processes, including but not limited to natural selection, both in real life and as depicted in *Star Trek*.

EVOLUTION, DEFINED THROUGH GENETIC CHANGE

ENT, Season 1, Episode 1, "Broken Bow"

Aliens of the species Suliban that can make themselves semitransparent invade the *Enterprise.* Doctor Phlox examines one alien's body and notes, "His DNA is Suliban but his anatomy has been altered. . . . This man was the recipient of some very sophisticated genetic engineering." The captain questions further, "It's not in their genome?" Doctor Phlox replies, "No, certainly not. The Suliban are no more evolved than humans." In a later encounter, another Suliban explains, "The price of evolution was too high. . . . Some of my people are so anxious to improve themselves that they've lost perspective."

The link between genetics and evolution is fundamentally important, and understanding this connection greatly increases one's appreciation of how evolution occurs.[1] Many writers simplistically define evolution as "change through time." This definition works in a nonscientific sense—we can say that our thinking on a controversial subject has "evolved." However, this definition is inadequate in biology, because it fails to capture the importance of genetics. We did not "evolve" from babies, nor do frogs evolve from tadpoles. Babies and tadpoles certainly change over time to appear physically different, but humans develop, rather than evolve, from babies into adults, and tadpoles develop, or metamorphose, into frogs. Despite physical changes over time, individual organisms do not "evolve" in a biological sense.

Instead, evolution occurs when the composition of a group (such as a population or an entire species) changes over generations. Evolutionary changes need not be dramatic nor even have an obvious physical manifestation. For example, 10,000 years ago, European human adults were nearly all lactose intolerant, experiencing gastrointestinal distress shortly after consuming any milk products. Today, the vast majority of European human adults regularly enjoy milk and milk products without ill effect. The difference in lactose intolerance between 10,000 years ago and today comes from the increase in abundance of a specific genetic variant that allows adults to digest lactose. Hence, the genetic makeup of the European human population changed over generations, meaning European humans "evolved" in the past few thousand years. The change is subtle; European humans before 10,000 years ago were still very much "human," and may not have had any difference in outward physical appearance from modern European humans. Note again that the word "evolution" here refers to the genetic composition of the population as a whole, not to any changes in specific individuals.

Importantly, these changes in composition need to be "inherited" to be considered evolutionary. Imagine a gardener has two sets of plants of the same species. The gardener gives the first set a lot of sun and water, while the second gets very little. When we observe that the first set is taller, we do not say it "evolved" to be taller or that there has been evolution in the overall group. The difference between these two sets is most likely a by-product of the environmental conditions that they encountered, not something inherited in the sets of plants. This distinction can be illustrated by reversing how the plants are treated—if one grows offspring from the short plants in the high sun and water environment, they may well grow to be just as tall as the tall plants. Evolution, therefore, requires inherited (transmitted from parent to offspring) changes over generations in a population, and the garden example above does not depict evolutionary change. As discussed in the preceding chapter, since epigenetic changes are generally not inherited over multiple generations, heritable changes driving evolution are more likely to be genetic than epigenetic.

The field of study called "population genetics" defines evolution as change in the abundance of particular gene variants over time in a population of a species. When a new mutation first arises in one individual, it is necessarily rare in the population. If the individual bearing the new mutation produces more offspring than average, then the mutation will become more abundant in the population in the next generation than in the present generation. If the individual bearing the new mutation produces fewer offspring than average or dies without reproducing, the mutation will be less frequent or even absent in the next generation. Both examples represent evolutionary changes over one generation. Forces driving these types of changes are the essence of the evolutionary process and the focus of the present chapter. Changes in multiple genes across a population can lead to cumulative effects on appearance or even species identity. For instance, humans differ from chimpanzees in a bit more than 1% of their genomes' DNA sequences. These differences reflect the accumulation of evolutionary genetic changes at multiple genes that happened in human populations and in chimpanzee populations over the last few million years.

The *Star Trek* series use the words "evolution" and "evolutionary" several times, though not always accurately. As a positive example, John in *TNG* "Transfigurations" describes his species being on the verge of an evolutionary change. John and other individuals have a mutation that eventually allows them to metamorphose into a glowing form that can fly.* John anticipates an

---

* The dialogue of the episode is unclear whether individuals inherit these "mutations" or whether these are mutations arising within these people's bodies. I will assume the former even though the episode seems to imply the latter.

evolutionary change wherein more members of his species will have this mutation and associated ability. Assuming this mutation is inherited by a larger fraction of his species over generations, his depiction is arguably fair that this change is "evolutionary." However, John also suggests that others (implying others already alive) can "join him," which is harder to explain in genetic terms.

The *ENT* "Broken Bow" discussion above regarding the Suliban merits some consideration. First, the episode discussion is correct that an evolutionary change in a natural population may stem from a genetic manipulation. For example, farmers worry about the spread of artificial genetic modifications into natural populations associated with the perceived potential for genetically modified (GM) crops to breed back with wild plants (or organic farms) nearby. This concern causes farmers to go to great lengths to isolate their GM crops.

However, the "genetic engineering" discussed in this *ENT* example seems strangely devoid of actual genetic change. Doctor Phlox suggests the changes are "not in their genome." Indeed, the alterations seem to have been directly added to their bodies, akin to giving someone an artificial heart. Such alterations would not be inherited; the recipient of an artificial heart does not have children with artificial hearts. As such, the changes cannot be "evolutionary," despite what the Suliban notes. However, it is possible that these Suliban individuals received tissues that were laboratory grown (like a skin graft), and the tissues were themselves genetically engineered to produce the enhanced abilities. While still not "evolutionary" since not passed on to offspring, such a possibility would explain how some of these enhancements could be removed from individuals later (*ENT* "Cold Front")—the genetically engineered tissues (eye retinas, in this case) were simply removed. Though removal of eye retina may not have felt all that "simple" to the poor individual involved.

NATURAL SELECTION: DIRECTIONAL GENETIC CHANGE

TNG, Season 3, Episode 1, "Evolution"

Young Ensign Wesley Crusher accidentally releases two self-replicating microscopic robots called "nanites" on the *USS Enterprise-D*. The nanites behave and reproduce based on information stored in gigabytes of mechanical computer memory within each robot.[†] After the ship starts to have severe malfunctions, Ensign Crusher and the crew deduce that the nanites have multiplied, "evolved," and caused the malfunctions. The nanites eventually become self-aware and able to communicate and interact with the ship's crew.

While mutations first come about due to random DNA copy errors, the fate of these new mutations may be quite nonrandom. The simplest case is when a mutation is disadvantageous to the organism bearing it. If a new mutation arising in one individual is both dominant (wherein one variant is sufficient to cause a complete manifestation; see chapter 3) and lethal (i.e., sufficient to cause death before the age of reproduction), then the mutation is necessarily lost from the population before the next generation. Scientists would say that natural selection eliminated the bad mutation from the population after one generation. This predicted outcome is robust to changes in the specific assumptions above, too. For example, if the new mutation is detrimental without being lethal (e.g., causing the person carrying it to have fewer offspring), or if the mutation is not dominant, the new mutation is still very likely to be eventually eliminated

---

† This episode first aired in 1989, when "gigabytes" of memory sounded large.

from the population, but the process may take more than a single generation. This process is common even in humans; recent studies have estimated that each human inherits on average the equivalent of two to four new detrimental mutations each generation,[2] most of which eventually get lost from the population over generations. Individuals that inherit more and/or worse mutations—the manifestation of which we may call a "genetic disorder," like hemophilia—are more likely to die before reproducing or to have fewer offspring.

While getting rid of detrimental mutations is a fundamentally important part of evolution, most people conceive of evolution by natural selection as the spread of advantageous new mutations across a population or species. Evolution by natural selection is often portrayed as part of a "struggle for existence," which it may sometimes be, but its action merely requires that some genetic variants allow the bearer to produce more offspring than others. Let us use the fictional *Star Trek* nanites from the episode above as an example. Ensign Crusher released two nanites that were likely virtually identical. Let us assume that nanites replicate hourly and that one of these nanites had a "mutation in its genome" (in this case, a random alteration in its memory file) that allowed it to self-replicate twice as fast as the other. For simplicity, we also assume that none of the nanites die during the observation period, which is unrealistic in a biological system over large numbers of generations but depicts what may occur over shorter periods reasonably well. To make natural selection easier to observe, let us say the nanite that reproduces once per hour is white in color and the nanite with the mutation causing it to reproduce twice per hour is dark in color (see figure 8). Finally, let us assume the nanite form is hereditary and they replicate "in-kind": white forms make more white forms, while dark forms make more dark forms. After one hour, or one

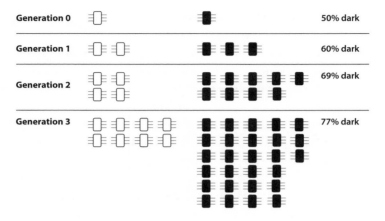

Generation 0		50% dark
Generation 1		60% dark
Generation 2		69% dark
Generation 3		77% dark

Figure 8. Increase in abundance of the dark nanite form in the population by natural selection.

generation, the white nanite would have one offspring, resulting in two white nanites. The dark nanite would have two offspring, resulting in three dark nanites—the parent plus two offspring. The population will have gone from 50% dark nanites to 60% dark nanites (three of five nanites). If we iterate this process over more generations, as shown in figure 8, the dark nanites make up more and more of the population, until eventually the vast majority of nanites one observes will be dark. Eventually, the original white form is evolutionarily equivalent to a "detrimental mutation" or "genetic disorder" with respect to the population: individuals of the white form have fewer offspring, and the relative abundance of the white form becomes rarer each generation until perhaps eventually being lost altogether.

With this example, I want to stress two points. First, natural selection is a "mathematical inevitability" if there is inherited variation that affects survival or number of offspring. The inherited variation in this fictional case was in the computer-programmed, self-replicating form in nanites, but it would work just as well as a genetic variant affecting claw muscle strength in crabs. Natural selection will inevitably (in the absence of other forces) result in evolutionary change in the genetic makeup of a population. Second, while Darwin and others emphasized the "struggle for existence,"[3] no struggle is necessary for natural selection to operate. Using the example in figure 8, we observe the gradual reduction in abundance of the white form relative to the dark form. However, if we look at the left half of the figure by itself, the white form is increasing in absolute numbers. If the dark form did not exist for comparison in the population, we might say the white form is doing quite well. Simply put, natural selection reduces the relative abundance of the "less fit" form, but less fit need not mean unfit.

The second point can be illustrated with a case of natural selection in humans. As mentioned previously, the ability to digest lactose as adults is evolutionarily new in European humans, and genetic data from ancient skeletons confirms that European human adults were nearly all lactose intolerant thousands of years ago.[4] Humans began to domesticate sheep, goats, and cattle around 10,000 years ago, providing access to milk and milk products. While all children can consume milk products safely, the (many) lactose intolerant adults at the time were unable to do so without pain and bloating. However, around 7,500 years ago, a mutation began to spread that allowed European adults to consume milk products safely.[5] This made it possible for those bearing the mutation to have access to more nutrition. This extra access to nutrition made them more likely to have more children (who also bore the mutation), and today most Europeans are lactose tolerant. Hence, the "mutant" form (able to consume lactose) is now considered by most people to be normal. Intriguingly, the same process of natural selection for lactose tolerance happened in ancient African populations thousands of years ago, but the process involved mutations different from the one that spread in Europeans. Still, just like the mutation in European populations, the mutations that spread in African populations also "cured" lactose intolerance from most of the populations.[6] Again, the ancient lactose intolerant humans were able to survive and reproduce for thousands of years before the mutations spread, so while they were "less fit" than the new tolerant forms on average, they were not "unfit."

I have stressed natural selection affecting change in humans here, but humans are also agents imposing selection on other species. Humans domesticated the mustard species *Brassica oleracea*, and we have applied selection to it for a few thousand years to create new, familiar vegetables such as cabbage, kale, broccoli, cauliflower, kohlrabi, and brussels sprouts. Humans created these vegetables through a process analogous to what happens with natural selection—we picked individuals having (genetically based) desirable traits and allowed those to reproduce disproportionately. Modern scientists have identified some of the gene variants associated with these forms,[7] but these were obviously unknown over most of the last few thousand years as selection occurred. The same type of selection has been applied to create dog or cat pet varieties we know and love, as well as cow breeds that we use in different ways.

Sometimes the effects of humans are less intentional, though. The antibiotic penicillin was first mass-produced in the early 1940s, and hailed as a "wonder drug" for treating infections. Very shortly thereafter, resistant strains of the infectious *Staphylococcus* (often called "Staph") bacteria were found. Presumably, both rare resistant and abundant nonresistant strains always existed, but when penicillin was introduced, nonresistant strains were killed, and resistant ones became more common. This cycle has been repeated with many other antibiotics, including methicillin ("MRSA" is the methicillin-resistant form of the bacteria),[8] leading to today when very few antibiotics are still effective in all known cases. Further, our long-term use of antibiotic soaps—for example, those containing triclosan—and their resultant introduction into wastewater has selected for resistant bacteria downstream of runoffs in the environment.[9] Microbes are not the only pests affected in this way—for example, insecticide resistant head lice are now extremely common and posing a major challenge for treatment.[10]

Returning to *Star Trek*, this process of new "mutations"—accidental edits in the computer code affecting survival or reproduction—introduced each generation and natural selection spreading the favorable mutations may explain the rapid evolution of the nanites in the *TNG* episode "Evolution." The advancement would be iterative: a new mutation may have arisen that allowed a nanite to harness energy better (thus increasing reproduction), then later another to promote better avoidance of energy discharges (thus increasing survival), then later another to increase processor operations, etc. As with genetic mutations, presumably most random changes in code would be detrimental rather than advantageous, but bearers of detrimental mutations would just disappear from the population. Bad mutations are quickly lost (or kept exceedingly rare), while good mutations quickly become abundant in the population. This process would repeat with each mutation and lead to iterative improvements, potentially even including the ability to communicate (assuming such an ability provided a survival/reproduction advantage). The episode mentioned the nanites bearing a "collective intelligence" a few times, but natural selection does not require intelligence or forethought. Again, natural selection is mathematically inevitable if there is inherited variation that affects survival or number of offspring.

One other noteworthy example of evolution by natural selection in *Star Trek* was featured in the *TOS* episode "Omega Glory." The *USS Enterprise* and *USS Exeter* crews meet humanoids on the planet Omega Four who live to be hundreds or thousands of years old. While the officers initially assume something in the environment leads to this long life, Doctor McCoy eventually shows the long life to be a product of "survival of the fittest"—an expression used to describe natural selection. Generations before the arrival of Earth starships, humanoids on this planet fought an intense biological war using viruses, and these viruses persisted in the environment. The people who survived this war were the subset who had genetic "superior resistance." After a few iterations of mutations arising that confer still-greater resistance to the viruses and selection making these mutations more common, the present-day descendants now enjoy unusually long life as a result. In general, older people are more prone to die of viral infections than the young,[11] so it does follow that acquiring greater immunity to viral infection may extend lifespan. I am skeptical that such an extension could lead to people living thousands of years; aging is also associated with factors unrelated to pressure from parasites, such as the age-related loss of repetitive DNA sequences called "telomeres" during cell division.[12,‡] However, perhaps the biology of Omega Four aliens is different enough from that of humans that viral parasites are a more significant cause of age-related death, and other factors causing human death do not apply to them.

‡   Impressively, telomere shortening with age was mentioned in passing in *ENT* "The Forge," but I will not go into detail on it in this book.

NATURAL SELECTION: TRAIT CHANGES OVER LONG TIME SCALES AND CONVERGENCE TO SIMILARITY IN FORM

TOS, Season 3, Episode 3, "The Paradise Syndrome"
Captain Kirk, Mr. Spock, and Doctor McCoy arrive on a planet bearing very familiar vegetation. They see what appear to be pine trees and smell something resembling honeysuckle and orange blossom. Captain Kirk notes, "It's unbelievable. Growth exactly like that of Earth on a planet half a galaxy away. What are the odds on such duplication?" Science Officer Spock replies, "Astronomical, Captain. The relative size, age, and composition of this planet makes it highly improbable that it would evolve similarly to Earth in any way."

Some of the public's resistance to accepting evolution comes from the perception of "design" in life.[13] Many organisms appear ideally suited to the environments in which they live or have amazing abilities. For example, birds have lightweight yet dense, stiff bones facilitating flight.[14] Mantis shrimp (figure 9) have a hammer-like arm that can break open snail shells for food. Despite these shrimp being only a few inches long, they can strike with their arm underwater at an impressive maximum speed of 69 miles per hour, while also increasing the impact force by leveraging a brief vacuum bubble they create in the water.[15] Reading any natural history book or watching Animal Planet on the television yields many, many more examples of amazing feats organisms accomplish through their structures and behaviors. Many are so amazing that some people may infer the organisms must have been "designed" to achieve these goals. Indeed, many

Figure 9. Nineteenth-century drawing of a mantis shrimp. *Source*: R. A. Lydekker (1896). *The Royal Natural History*, volume 6. Frederick Warne.

engineering projects invoke "biologically inspired" systems (e.g., artificial drones copying insect wing motions, even on such nonbiological ventures as space missions),[16] so we humans are actively "designing" systems based on ones we observe in nature.

In reality, serial rounds of mutation and natural selection create the appearance of design. Natural selection is neither directed nor goal oriented—it merely allows the spread of any genetic variant that causes an organism to have more offspring. Possessing such favored genetic variants may make an organism better at flying or increase its strength in securing prey or something altogether different. Specific kinds of features may be favored by natural selection in particular environments. For instance, being white in color would be advantageous for animals in the arctic because they would be better camouflaged. Stream-lined, elongate bodies and powerful fins or flippers are advanta-geous for sea animals. Any mutation that even slightly improves these forms to fit better with their environment increases (even if infinitesimally) the average rate of reproduction of an organ-ism and will spread by natural selection. Over millions of years, many such mutations will arise and spread, and the forms may become strikingly different, causing gross changes in body form that we often refer to as "macroevolutionary."

One challenge to the acceptance of evolution by the pub-lic is an underappreciation for time scales. If we say an event happened 40,000 years ago, that feels like a very long time ago, far longer than a human lifetime. If we say an event happened 400 million years ago, that again feels like a very long time ago, far longer than a human lifetime. However, 400 million years is 10,000 times longer than 40,000 years, yet we cannot readily conceive of the difference between these time scales. And a lot of evolutionary change in appearance can happen in a much shorter time. Let me illustrate this point with two examples of selection by humans. Domestic dogs diverged from wolves less than 40,000 years ago.[17] If we conservatively assume that separate dog breeds began to be selected at that time (which is unrealistically conservative, as many dog breeds were just se-lected within the past 1,000 years), then we conclude selection can make forms as dramatically different as Chihuahuas, Chow Chows, Beagles (like Porthos in *ENT*), Golden Retrievers, and Great Danes in that much time. These forms differ dramatically in shape, form, and behavior. Now, imagine animals potentially 10,000 times more different from each other in any dimension of shape, form, or behavior than Golden Retrievers and Chihua-huas. We can do the same calculation with plants; as mentioned, humans domesticated *Brassica oleracea* around 2,000 years ago and selected it into the distinct forms we now call kale, collard greens, broccoli, cauliflower, etc. If selection can make vegetables appear so different in just ~2,000 years, imagine how different they might be in 400 million years (or 200,000 times longer), which is still well after the origin of land plants.[18]

Natural selection optimizes an organism to its environment over time. With enough time in the same type of environment, species that are very distantly related can evolve by natural selection to appear similar in form: a process called convergent evolution. For instance, dolphins and sharks have the same overall streamlined body with a triangular fin on the back and two fins on the sides. Yet dolphins have land-dwelling mammalian ancestors from within the past 50 million years,[19] whereas the common ancestor of sharks and dolphins may have lived almost 10 times as many years earlier. Sugar gliders and flying squirrels also appear strikingly similar in structure and size, each with flaps of skin stretched between their legs and used for gliding, but they are only very distantly related to each other, and each is far more closely related to various nongliding species. In both of the pairs mentioned here, the similarity in form belies many differences between these pairs of species. Focusing on the first pair, sharks have skeletons made of cartilage while dolphins have bony skeletons, sharks get oxygen from gills under water while dolphins breathe air through their blowholes, sharks do not nurse their young but dolphins do, etc. The facets of appearance that have "converged" by natural selection over time are ones specific to taking advantage of a particular environment, and allow those species to have more offspring in that environment.

Similarities in form via convergent evolution need not involve the same mutations arising in the different species or populations. For example, as mentioned earlier, European human populations and African human populations have different genetic mutations that allow adults to digest lactose. Although the product of natural selection was the same—the ability to digest lactose—a different mutational road was taken to achieve it. Sometimes different structures are co-opted by mutations and selection to make similar products. For instance, pandas and humans both have fingers and thumbs, but the "thumb" pandas use has only one bone and develops from the wristbone as a sixth finger.[20] Still, it serves the same grasping function as our thumb and appears superficially similar.

Now let us return to *Star Trek*. In chapter 2, I discussed the hypothesis from *TNG* "The Chase" that humanoids had evolved independently on different planets. I emphasized the importance of random events—for example, environmental changes associated with volcanic activity and a major asteroid impact—as a reason to not expect the same outcome. However, even if we assumed essentially the same environment, might we expect such similar forms via convergent evolution? Again, convergence can be impressive on specific traits but often belies many more subtle differences. When we think of many of the humanoid species in *Star Trek*—humans, Vulcans, Betazoids, El-Aurians, Bajorans, Trill, Ocampans, etc.—they really do look exactly the same beyond some subtleties of ear, eyebrow, or nose shape. A few species are somewhat more different (e.g., Ferengi, Klingons outside of those depicted in *TOS*), but the convergence is far more overt than just overall body form. In fact, they are so similar physiologically and genetically that many of these species can even successfully interbreed—a point to which I will return in the next chapter. Dolphins and sharks do not look nearly this similar, nor would (or could) they ever interbreed.

In the example at the start of this section, Mr. Spock rejects the idea of convergence of plant life based on "the relative size, age, and composition of this planet" relative to Earth. I argue that the qualifier is unnecessary; even if the planet were exactly the same size, age, and composition of Earth, though some convergence in life forms might occur (e.g., perhaps in shape of swimming organisms), seeing a forest of trees that look and smell exactly like multiple species on Earth would also be "astronomically" unlikely.

There was one explicit mention of evolutionary "convergence" in *Star Trek*. In *VOY* "Distant Origin," the genetic similarity between the Voth aliens and Earth organisms (including humans) was erroneously dismissed by a scientist as "a result of random convergence." Because he specifies this convergence as "random," he is discounting that natural selection drove the genetic similarity and instead still uses the term to suggest that the similarity is not a result of recent shared ancestors. Random processes can lead to various evolutionary changes including convergence,[21] and such processes are the topic of the next section.

EVOLUTION BY RANDOM CHANCE AND INBREEDING IN SMALL POPULATIONS

DS9, Season 5, Episode 22, "Children of Time"

The *USS Defiant* crew comes upon a planet whose inhabitants are actually their own descendants; in an alternate timeline, the *Defiant* crew accidentally went back 200 years in time and crashed the ship, and the 48 survivors settled on this planet. Now, 200 years later, there are almost 8,000 people living on the planet, descended from the original 48.

Although many people (including many biologists) perceive evolution as a continued process of "improvement" driven by natural selection, much evolutionary change can be quite random. Some of this randomness comes about because some individuals in virtually every species by chance reproduce more or less than others; for example, an individual with gene variants conferring good health may still have an accident and die before reproducing. Other elements of this randomness come about because of how variants are transmitted via genetics. For example, if I have one genetic variant for lactose intolerance and one for tolerance (analogous to the antennae example in chapter 3/figure 7), it is possible that I could transmit the intolerance variant to both of my children without ever passing along the tolerance one. Chance events can have large cumulative effects on evolutionary change in the genetic composition of populations, even beyond major environmental catastrophes (as discussed in chapter 2). Random change in the abundance of genetic variants in a population over generations is called "genetic drift."

Random chance has especially large effects when there are very few individuals in the next generation. Small samples can cause large random chance deviations in the percentages of specific genetic variants—just as flipping a fair coin 100 times results in approximately 50% heads and 50% tails, flipping a fair coin just twice may easily result in two (100%) heads or two (100%) tails. I will illustrate this concept with sampling from a population. Imagine that aliens were to take people from the city of San Francisco (the location of Starfleet headquarters in *Star Trek*) and start a series of descendant populations with the inhabitants. For mathematical simplicity, let us pretend that humans reproduce asexually by replicating between zero and two copies of themselves and that people die immediately after they reproduce. Let us further assume that race is genetic and perfectly inherited, even though the reality is that races are not genetically homogenous and lack clear-cut genetic boundaries. Finally, we assume that no race has a biological difference causing it to reproduce more. Today's racial demographics for San Francisco are roughly 48% Caucasian, 33% Asian, 6% African American, and 13% other. Now, imagine that the aliens place 100,000 people at random into population 1. Probabilistically, we expect the racial demographics of population 1 to closely reflect that of San Francisco. After those transplanted people reproduce, the percentages will still be pretty similar in the next generation. The aliens place only 50 people at random into population 2. By chance, population 2 will probably not have exactly the same racial demographics as San Francisco—for instance, population 2 may easily be 40% Caucasian, 38% Asian, 10% African American, and 12% other. In the next generation, the fractions may become more different still from the original population due to chance. Finally, the aliens place only 4 people into population 3. It is now impossible to have even close to the same racial demographic as San Francisco—it may be 1 (25%) Caucasian, 2 (50%) Asians, and 1 (25%) other, hence losing one category altogether. In the next generation, another category may be lost due to chance if, for example, the one Caucasian human fails to reproduce.

Small populations can also lead to inbreeding: individuals mating and reproducing with close relatives. In a population of 100,000 people, it may be easy to find a mate who is very distantly related, but in a population of 50 people, it may be that the only available mate is a first cousin, for example. We often associate the idea of inbreeding with cultural taboos and genetic disorders. Inbreeding causes disorders due to the union of two properties: (1) most individuals harbor mutations with bad effects (see chapter 3), but most of us harbor *different* bad-effect mutations, and (2) most mutations with bad effects are recessive (see chapter 3), meaning that two copies of the genetic variant (one inherited from the mother and one from the father) are needed for the bad effect to manifest itself.[22] Most individuals have just one copy of any particular bad-effect mutations they harbor. With inbreeding, however, the mother and father are closely related, thereby likely having the same bad-effect mutations. When they reproduce, they may both transmit the same bad-effect mutation to the offspring, causing the offspring to have the genetic disorder.

How did genetic drift and inbreeding play out in *Star Trek*? The term "genetic drift" was used twice in the series but inaccurately both times. In *TNG* "Second Chances" and in *DS9* "A Man Alone," the respective doctors look at a single individual's genetic information and claim to look for "genetic drift." However, genetic drift is a process that occurs at the level of populations—random changes in the relative abundance of genetic variants over generations—not in single individuals. The term does sound interesting and technical, though, so I understand why the writers may have seized upon it.

Inbreeding was not discussed specifically but it surely happened in many instances in the series. One extreme version was featured in the movie *Star Trek IV: The Voyage Home*. Two humpback whales (George and Gracie) were brought forward in time from the 1980s to the twenty-third century to repopulate the species. Repopulating the species would necessarily involve George and Gracie's offspring having children with their siblings. George and Gracie's offspring would likely have inherited some of the same bad-effect mutations from their parents, so it's likely that some of their children will inherit two copies of some mutations and suffer ill effects. Even though the mission to bring the whales forward in time was successful in the movie, it is unclear whether humpbacks would persist as a species very long in the twenty-third century without heavy interventions, given the negative effects of extreme inbreeding.

The *DS9* example mentioned at the start of this section depicts a less extreme example of genetic drift and/or inbreeding. The new population on Gaia was founded with the 48 surviving members of the *USS Defiant*. This is a small population initially, so some evolutionary changes via genetic drift are likely. For example, one of the survivors was the chief engineer, Miles O'Brien, who is of Irish descent. The Irish have a higher-than-average likelihood of being carriers for mutations causing Tay–Sachs disease (one estimate was 1/25 among the Irish, compared with 1/150 in the general European population).[23] The episode noted that the population of Gaia had "a lot of" people named Molly descended from the O'Brien lineage, potentially implying that some genetic drift made O'Brien's descendants more common than others' descendants. If Miles O'Brien happened to be a carrier of a Tay–Sachs-causing mutation, then the population of Gaia could have an unusually high incidence of Tay–Sachs disease overall. While this example is hypothetical even within the context of the episode, an analogous situation actually occurred on the island of Mauritius: around 1800, a French immigrant to the island carried the disease Huntington's chorea and had many descendants, and the disease is now roughly 10 times more abundant among the Europeans on the island (1/2166)[24] than among Europeans worldwide.[25]

In both the case of Huntington's chorea on Mauritius and potentially that of Tay–Sachs on Gaia in *DS9*, genetic drift causes a mutation that is unfavored by natural selection to still spread within a population, even though the population eventually becomes quite large (almost 8,000 in the case of Gaia). Genetic drift can be more powerful than (and oppose) natural selection if population sizes are quite small even for a few generations.

However, changes caused by genetic drift are random in direction: drift is equally likely to make an unfavorable mutation increase or decrease in abundance in any particular generation.

PROGRESS AND PROSPECTS STUDYING EVOLUTIONARY CHANGE

This chapter has briefly discussed a few evolutionary forces—natural selection, genetic drift, and inbreeding both in general and as depicted in *Star Trek*—that contribute to genetic changes over time in natural populations of most species. Population geneticists sample DNA from many individuals of many species (including but not limited to humans or even animals) and apply statistical models to the DNA sequences to infer what evolutionary forces may have been operating. New technologies associated with faster and cheaper DNA sequencing have accelerated many of these efforts; the estimated cost for generating the initial draft human genome sequence was approximately $300 million, while generating new human genome sequences can now be accomplished for barely $1,000. These genetic studies can generate hypotheses on the focal species that can be tested by characterizing genetic variation in natural populations or by direct experimental manipulation in the laboratory.

Partially as a result of such technologies alongside new computational approaches, population genetics has produced many new insights in the last decade. Mutation rates across generations have been estimated to unprecedented precision, and researchers have even shown that older fathers transmit more mutations than young fathers to their children.[26] Statistical analyses of human genome sequences inferred that approximately 6% of the human genome has experienced natural selection against bad mutations.[27] Researchers have found specific genetic changes that spread via natural selection in particular environments, such as ones conferring the ability to get enough oxygen at high altitudes in ethnic Tibetans, Andeans, and high-altitude Ethiopians.[28] Numerous studies have also estimated changes in human population sizes over history,[29] as well as inbreeding within populations and its potential role in disease.[30] While I describe studies of human populations here, population genetic analyses have been conducted on a wide range of organisms, allowing for exciting evolutionary comparisons.

Still, even with these advances, long-term evolution is far from "predictable." As mentioned earlier, distinct species can converge in overall form when living in similar environments, but even these convergent forms retain many differences. Additionally, genetic drift causes evolutionary changes in random directions, and specific mutations may arise in one species but not another, allowing adaptation to proceed through entirely distinct means. Further, environments often change unpredictably over the long term, so we cannot even hypothesize what forms may be favored by natural selection. Hence, while we may estimate "how much" change we expect may occur over a short time scale, the magnitude or direction of change in the long term remains obscure.

In this regard, long-term evolution is often depicted as unrealistically predictable in science fiction, including in *Star Trek*. There are many references to "the next step in our evolution" (*Star Trek: The Motion Picture*), species being "on the verge of a wondrous evolutionary change" (*TNG* "Transfigurations"), species anticipating an "evolutionary awakening" (*ENT* "Dear Doctor"), etc. If a particular genetic variant or inherited trait has been observed and unambiguously contributes to greater survival or increased reproduction, then it is logical to infer that it is likely to spread by natural selection, causing an evolutionary change. However, such a variant or trait may also be lost from the species by genetic drift, or perhaps a change in the environment may make it less favorable in the near future. Things get vaguer still when we try to speculate what a species "will evolve into" on a time scale of tens or hundreds of millions of years. The origin of new life and its accelerated evolution associated with the Genesis device in *Star Trek II: The Wrath of Khan* repeatedly resulting in familiar plants such as various trees and elephant ear plants—though, curiously, no animal life—is wholly unrealistic (see also the earlier discussion of the importance of random events in chapter 2).

While we may not be able to predict what a species will evolve into over time, the enormous diversity of species on our planet proves that new species do form from old species. Hence, new species formation is a common outcome of evolution on our planet. The next chapter explores characteristics of new species as well as how mate preferences evolve. This discussion is particularly relevant to *Star Trek*, in which interbreeding among alien species seems to have few, if any, barriers.

CHAPTER 5

SEX, REPRODUCTION, AND THE MAKING OF NEW SPECIES

Star Trek and other science fiction series depict "love" and sexual reproduction as universal. While sexual reproduction is taken for granted by us as humans, researchers have spent decades trying to understand its high prevalence across life on our planet. One might infer that asexual reproduction would be preferable. For example, sexual reproduction requires that two individuals participate in the production of one offspring. Further, offspring only get half of each parent's genetic material, resulting in a so-called "two-fold cost of sex." Further, there may be added costs to reproducing with another individual, including time and energy used seeking a mate and potential disease transmission. For sexual reproduction to be advantageous and spread via natural selection, sexual couples should produce on average more than twice as many offspring as if each participant reproduced alone.

However, science fiction series, including *Star Trek*, take sexual reproduction further still. They often depict members of different species falling in love and/or reproducing with each other.

One can break down exploring interspecies reproduction into a series of questions, addressed using species on Earth as well as aliens in *Star Trek*. Why do so many organisms on Earth have sexual reproduction? Why do some individuals contribute to the reproduction of others rather than reproducing themselves? How is sexual reproduction associated with the process of species formation? What are problems faced by species hybrids? I will explore each of these questions in this chapter.

WHY SEX?

TOS, Season 2, Episode 15, "The Trouble with Tribbles"
(one of the most entertaining episodes across all the *Star Trek* series)

While on a space station, a salesman gives Lieutenant Uhura a small furry animal called a "tribble." She takes the tribble back to the *USS Enterprise*, and after a few days, there are tribbles all over both the space station and the ship. Captain Kirk asks how the tribbles got to be so abundant. Doctor McCoy explains, "The nearest thing I can figure out is they're born pregnant, which seems to be quite a time saver. . . . And from my observations, it seems they're bisexual, reproducing at will. And, brother, have they got a lot of will."

A fundamental feature of reproducing "sexually" with other individuals is that the act typically generates genetic combinations in the offspring that are distinct from those of the parents. In contrast, asexual reproduction, in its most basic form, produces a genetic copy of the mother. Many animal and plant species exhibit sexual reproduction as we know it: union of paternal and maternal genetic information via union of a sperm and egg (or pollen and seed). However, even bacterial species we perceive to be asexually dividing from one cell into two identical cells also sometimes transfer some DNA between individuals or take in DNA from their surrounding environments (see Appendix), thereby also producing new combinations.[1] As such, the process of genetic exchange and formation of new combinations, fundamental but not exclusive to sexual reproduction as we know it, happens very broadly in organisms on Earth.

Researchers have proposed many theories to explain the widespread occurrence of genetic exchange in general and sexual reproduction in particular, but many of these theories fall into two general categories.[2] One category of these explanations is simply that an advantage exists to producing offspring not genetically identical to each other or to the parents. As mentioned in the preceding chapter, we cannot anticipate what traits will be advantageous in the future. Even if an organism has a genome sequence that is optimal for the present environment, the environment in future generations may be quite different. If one continues to produce identical copies of oneself but the environment changes, then all descendants may be lost. We can think of this with a "lottery ticket" analogy—if you do not know what the winning lottery numbers will be, it makes more sense to pick a suite of different numbers for each ticket than to get many tickets all bearing the same number. Importantly, producing this diversity is not a conscious decision by organisms to ensure survival of their offspring. Instead, this explanation for sexual reproduction posits that lineages that produce varied offspring tend to have at least some descendants persist over long evolutionary times.

The other category of explanation has to do with eliminating bad mutations. As discussed in the preceding chapters, new mutations that reduce health or reproduction arise in almost every generation. Imagine a lineage of individuals that reproduce without any sort of mating or exchange. The offspring will inherit the bad mutations that their parents possessed, and they will also get new bad mutations that arise. This process will repeat every generation. While a few good mutations may also arise, the lineage will very likely go extinct eventually because its descendants will be loaded with so many bad mutations that individuals cannot reproduce. One can think of this like taking serial photocopies of an image. The first copy will be nearly identical, but as one subsequently makes "copies of copies," the image will eventually have quite poor quality. This process would also occur when making genetic copies (clones) from other clones repeatedly: serial degradation would occur if the process is imperfect.

Sexual reproduction and other forms of exchange with other individuals reduce this "loading" of more and more bad mutations. While some descendants inherit the bad mutations from previous generations, other descendants inherit the nonmutant forms of these genes from other ancestors. This process leads to more variation among the offspring—some as loaded as the previous generation, some even more loaded than the previous generation (perhaps getting bad mutations from both parental lineages), and some that are less loaded. Natural selection then favors this last category, who thereby leave the most descendants. Returning to the photocopy analogy, if there were two copies made, and a subset of "offspring" could take the best elements from each copy, degradation would slow or stop.

Researchers have proposed other potential advantages to sexual reproduction and exchange,[3] and the balance of benefits and costs with other factors contributes to why we see some species exhibiting sexual reproduction and others not. Early studies suggested that asexual reproduction evolves occasionally from sexual ancestors but only persists for short evolutionary times, and they suggested asexuality is therefore disadvantaged relative to sexual reproduction over the long term. However, recent studies have questioned whether a general disadvantage exists.[4] For instance, a few species appear to have retained asexual reproduction for tens of millions of years.[5] Examples include some bdelloid rotifers, although these may occasionally have other forms of genetic exchange.[6] On the other hand, several species thought to be primarily or exclusively asexual appear to have occasional bouts of sexual reproduction or other forms of exchange.[7] Biologist John Maynard Smith once remarked that the "evolution of sex is the hardest problem in evolutionary biology,"[8] and it remains an area of active research.

The *Star Trek* series depict a few asexually reproducing species. For example, the Xindi insectoids are genderless and produce egg sacs asexually that are presumably clones of the parent (*ENT* "Hatchery"). Impressively, *Star Trek* directly acknowledged degradation from serial asexual reproduction in clones, and called it "replicative fading" in *TNG* "Up the Long Ladder." The Vorta in *DS9* are also genetically engineered clones. Consistent with a mutational process of some sort during clonal reproduction, clones of the same Vorta occasionally had very different personality traits (Weyoun-6 in *DS9* "Treachery, Faith, and the Great River"). However, unlike depictions of the Vorta, clones should not inherit conscious memories from previous versions. In all species we know, memories do not get encoded into genes and passed on through gametes. Based on inheritance in life forms with which we are familiar, clones would be no more likely than identical twins to share conscious memories.*

* *TNG* got this right in the context of the clone of Kahless in "Rightful Heir": the Klingons had to imprint specific memories so he would know some of Kahless's experiences, but these memories were obviously incomplete. However,

Interestingly, the tribbles mentioned at the start of this section would NOT be an example of asexual reproduction. Doctor McCoy explicitly mentions that they are "bisexual" and that they are born pregnant. The implication is that tribbles self-fertilize prior to birth. Self-fertilization is known in some plant and animal species. Importantly, however, self-fertilization does not result in a clone of the parent. Imagine an individual has one mutant copy of a gene and one normal copy of the gene. A clone or asexual offspring of that individual would necessarily also have one mutant copy of the gene and one normal copy of the gene. However, if that individual self-fertilizes, three possible types of offspring may emerge—one like the parent, one with two mutant copies of the gene, and one with two normal copies of the gene. As such, some tribble offspring will be born without some of the mutations of their mother. Given that self-fertilization is a rather extreme form of inbreeding, which has consequences discussed in chapter 4, the writers were wise to have tribbles produce so many offspring, some of which might be more likely to survive.

WHO REPRODUCES, AND HOW?

ENT, Season 2, Episode 22, "Cogenitor"

The *Enterprise* crew meet individuals of the Vissian species, and many crewmembers are intrigued by their unusual reproduction. Vissians can be traditional male or female,

---

VOY "Demon" got this wrong with the silver-blood DNA clones having the memories (and language) of the original *USS Voyager* crew that they copied. Similarly, *ENT* "Similitude" got this wrong in having Trip's clone bear his memories. This was particularly obvious when the sample (blood?) was drawn from Trip's neck, where no memory engrams would exist.

but 3% of their population are "cogenitors" who aid with impregnation. Doctor Phlox notes that there are other species who have more than two sexes and speculates that "the cogenitor provides an enzyme which facilitates conception." Details of this process were not presented, and most of the episode focused on the social treatment of cogenitors.

Even within Earth's animal kingdom, several species deviate from reproducing via genetic contributions from a male father and female mother. For example, the preceding section mentioned asexual reproduction by some animals such as bdelloid rotifers. As also mentioned in the preceding section, although sexual reproduction is quite common across animals, sometimes asexuality evolves from sexual species. In some cases where asexuality evolved recently, female animals still engage in copulation with males but without the male contributing any genes to the offspring. For example, one species of Amazon molly (a fish) is all female and mates with males of other similar species.[9] Fertilization in this species triggers development, but the males' genes are not included in the offspring: offspring are simply clones of the mother. Hence, the males are simply used by females of this species to stimulate their own reproduction.

Males of some other species can be as sneaky as the female Amazon mollies. In the freshwater Asian clam, fertilization occurs between a mother's egg and sperm bearing both copies of the father's genes, but the mother's genome is then eliminated from the developing embryo.[10] As such, the offspring develop into clones of the father. Scientists have observed similar processes in several other species, including a fish.[11]

Both sets of examples above still involve a distinct male and female (even if the genetic contributions are limited to one), but reproduction on Earth need not involve two consistently distinct sexes. Many species are hermaphroditic: bearing reproductive structures associated with both males and females. Some species, such as clown fish, can transition between being nonbreeders, males, and females during their lifetime. Clown fish live in structured groups wherein the female is the largest, the male is the second largest, and a few nonbreeding, gonadless individuals are also around.[12] When the female dies, the male becomes a female, and the largest nonbreeding individual becomes a male. (This process suggests that Marlin in *Finding Nemo* may have transitioned to a female after Coral died.) Individuals of many other species are simultaneous hermaphrodites. Most individuals of the nematode species *Caenorhabditis elegans*, often used in genetic and developmental biology research, are simultaneous hermaphrodites that can fertilize themselves, similar to the *Star Trek* tribbles mentioned above. However, 1% of the nematode population are males that can fertilize the hermaphrodites, thus introducing occasional genetic exchange in the species. Earthworms, in contrast, are hermaphrodites, but they pair with other earthworms (rather than self-fertilizing), leaving both individuals fertilized.

Less common on Earth are cases where more than two individuals participate in reproduction, similar to the *Star Trek* example at the start of this section. Social insects' (e.g., bees, ants) reproduction involves efforts by multiple types of individuals, including a female queen, male drones, and workers. Using honeybees as an example, the queen will mate with one or more drones. Unfertilized eggs develop into male drones (thereby having no father but having one copy of every gene from their mother), while fertilized eggs typically grow up as female worker bees. However, "parental care" is provided exclusively by the worker bees. Further, a subset of larvae from fertilized eggs are fed an excess of royal jelly by these worker bees, and these larvae develop into queens. While only the queen and drones provide genetic material to worker or queen-bee offspring, the worker bees are still critical to successful reproduction given their roles in caring for the queen and the offspring. Such "nest helpers," individuals who help others reproduce, have been observed in some bird and mammal species as well.

Helpers who fail to reproduce directly but who aid others' reproduction seem to present an evolutionary conundrum. One might presume that natural selection would not favor a system that requires extra individuals to participate in reproduction, and a selective advantage would exist for making only "reproducing" rather than "helper" types. A solution to this paradox was elaborated over 50 years ago (even before *TOS*): nonreproducing individuals aid the spread of their genes if they help close relatives to reproduce.[13] Natural selection favors traits in individuals that make their genes become more abundant in a population, irrespective of who transmits them to the next generation. For example, humans share 50% of their genes with their children and 25% of their genes with their nieces and nephews. If an individual can have two children without help (each having 50% of the parent's genes), or if that individual could instead assist a sibling to produce an additional five surviving nieces or nephews (each having 25% of the aunt's or uncle's genes), natural selection could favor the latter strategy. Correspondingly, nest helpers and worker bees tend to be siblings or offspring of the reproductive individuals, and their efforts thus facilitate transmission of their own genes.

Despite a few misstatements—for example, Captain Kirk saying, "The idea of male and female are [*sic*] universal constants, Cochrane. There's no doubt about it" in *TOS* "Metamorphosis"— the *Star Trek* series admirably consider alternative forms of reproduction. Tholians, for example, are said to "possess both male and female characteristics" (*ENT* "In a Mirror, Darkly, Part 1"). Several species are identified as "genderless" in some manner, such as the J'Naii in *TNG* ("The Outcast") and Morn in *DS9* ("If Wishes Were Horses"). The example at the start of this section regarding the "cogenitors" in *ENT* is analogous to the situation in social insects—while the workers are not directly involved in the act of reproduction, they are necessary for successful reproduction, including survival of the offspring. A cogenitor-like strategy could potentially evolve if, at least initially, the cogenitors preferentially help close relatives over nonrelatives, perhaps greatly increasing offspring production (maybe dramatically increasing fertilization success) or survival. Although the cogenitors are not directly reproducing, such a strategy applied to relatives would still facilitate the transmission of their own genes as well.

Examples of "more than two sexes" come up both in *Star Trek* and in the scientific literature, but what this means is often misinterpreted. For instance, the Doctor in *VOY* notes that the unusually named "species 8472" appears to have as many as five sexes.[†] This comment was not elaborated: does he mean that a single reproduction requires all five sexes to participate? Based on reproduction on Earth, I infer that he means a subset. For example, the microscopic freshwater species *Tetrahymena thermophila* has seven mating types (or "sexes"). Each can reproduce asexually or sexually with another individual, but sexual reproduction only works between individuals of different mating types. Despite the many sexes, it still only takes two individuals to tango. Moving to animals, one scientific study got a lot of popular press in 2016, with headlines like "The Sparrow with Four Sexes."[14] Rather than involving a strange reproductive ritual

---

† *VOY* "Someone to Watch Over Me."

with all four sexes together, this sparrow also has multiple genetic types, and only a subset can reproduce with each other. Like with *Tetrahymena thermophila*, each reproduction in these sparrows still involves only two individuals, and the same may be true for species 8472 in *Star Trek*.

Several other interesting and instructive examples came up in the *VOY* series that resemble mating tactics on Earth. The Taresians ("Favorite Son") manipulate males from other species into mating with them and then sacrifice them. Some fireflies on Earth exhibit "aggressive mimicry" associated with mating: *Photuris* females flash their lights, mimicking females of other firefly species, to attract those males and then eat them when they arrive.[15] The Kobali ("Ashes to Ashes") inject their DNA and "reanimate" corpses of other species. While nothing directly analogous happens in any Earth species, the DNA replacement process has elements similar to the freshwater Asian clam example above. Finally, the Ocampa in *VOY* present a bit of an evolutionary enigma. Unlike Vulcans, who have a ritual mating cycle every seven years (first mentioned in *TOS* "Amok Time"—I return to this case later), Ocampan females have one conception in their lifetimes ("Elogium"). For a species with two equally abundant sexes to survive, females must on average produce two surviving offspring. Some species do this by producing very large numbers of low-investment offspring of which very few survive, while others invest heavily in a few offspring to ensure their survival. This begs the question of how the Ocampan species persists if each female has one conception and produces one offspring. "Elogium"'s script strongly implied that a single offspring would result from character Kes's pregnancy, but perhaps some specific individuals produce a much larger number of offspring. Males in most sexual species on Earth produce highly variable numbers of offspring,[16] but females exhibit some variance as well. For example, in many insect species, larger females can produce more offspring than smaller ones,[17] so perhaps the limit to one offspring was associated with Kes's small size rather than being typical of all female Ocampans.

Assuming some individuals die, reproduction is obviously fundamental to the persistence of life in general and species in particular. However, reproduction is also fundamental to what keeps different species distinct. Individuals tend to mate and combine their genes with members of their own species, but successful mating and genetic exchange is less likely or absent with individuals of other species. Indeed, biologists often define separate species as groups that do not (or rarely) interbreed and are genetically distinct. In the next few sections, I will explore some kinds of traits that keep species genetically distinct.

PREVENTING INTERSPECIES MATING

DS9, Season 2, Episode 6, "Melora"
Ensign Melora Pazlar is a member of the Elaysian species, and comes from a low-gravity planet. Elaysians can glide in the low gravity of their planet, but they require wheelchairs or other assistance in an environment with Earth-like gravity. Ensign Pazlar becomes close to Doctor Bashir (a human) and later talks about her interest in Doctor Bashir to his friend Lieutenant Jadzia Dax. Ensign Pazlar notes with frustration, "Our species are just so different." Lieutenant Dax replies, "Since when has that ever stopped anybody? I knew a hydrogen-breathing Lothra who fell hopelessly in love with [someone who breathed oxygen]."

Species are often defined as groups of individuals that can interbreed successfully with each other but do not interbreed with other such groups.[18] "Successful" interbreeding in this context means both that individuals of a species will mate with other individuals of that same species and that offspring are produced that are also capable of reproducing (I will return to the offspring in the next section). Interspecies mating does happen in nature—one review estimated that at least 10% of animal species and 25% of plant species have individuals that mate with another species.[19] However, interspecies mating is rarer than that estimate implies. I interpret this figure to mean that 90% of animal species may have virtually 0% of individuals mate with other species, and most of the remaining 10% of species have had only small fraction of individuals do so. Hence, examining individual organisms rather than whole species, the probability of interspecies mating is likely far less than 10%. In this section, I explore reasons why interspecies mating is rare. For an interspecies mating to occur, the individuals need to both encounter one another at a time when breeding can happen and be willing and able to mate with the individual they encounter. As such, factors that affect whether interbreeding happens can involve space, time, or behavior.

Many spatial factors affect whether individuals mate. Circling back to a topic discussed in chapter 1, some of these factors may involve tolerances. Animals that live exclusively in hot environments are very unlikely to breed with animals that live exclusively in cold environments. Given much of life lives in water, we can also deduce that organisms requiring high salinity environments will not interbreed with those requiring low salinity environments. Such broad environmental factors prevent interbreeding between life forms that have adapted to them because the species are simply incapable of occurring together. However, spatial factors preventing interbreeding may occur on far finer spatial scales and involve behavioral preferences. For instance, the North American fruit fly *Rhagoletis pomonella* exhibits multiple genetically distinct forms, one breeding on apples and one breeding on hawthorn berries.[20] These trees are sometimes quite close to each other, certainly separated by distances that a fruit fly could fly. Flies of this species return to the same type of tree bearing the fruit on which they grew up, even if removed from their host and multiple trees of the other type are nearby, and they mate and lay eggs on this same type of fruit.[21] As such, the two forms almost never interbreed with each other and may be thus in the process of becoming new species. This example is striking because the species is thought to have first begun to breed on apple trees only in the past 160 years, so it has rapidly progressed toward becoming two distinct species over a few human generations.

Differences in timing of either mating behaviors or gamete release can also prevent species from interbreeding successfully. Mating or gamete release may occur in specific seasons of the year, and if two species differ in the timing of these mating seasons, then no interbreeding occurs. Sufferers of seasonal allergies are perhaps acutely aware of the seasonality associated with when particular kinds of pollen are released by grass or trees, for example. The separation may occur over much longer time scales than a year. For example, the cicada species *Magicicada tredecim* and *M. septendecim* have 13- and 17-year life cycles respectively. They spend those years burrowing around underground as juveniles, and they then only emerge and breed as adults for a few weeks at the end.[22] Both species appear to have evolved from a 13-year-cycling ancestor,[23] and their current timing distinction makes it so they rarely encounter the other species at breeding times. Differences in timing may also occur on much shorter time scales. Some tropical western Atlantic coral species (*Montastraea annularis* and *M. franksi*) release their sperm and eggs a few hours after sunset, and fertilization occurs to

produce offspring. However, these two species release their gametes at slightly different times of day, differing by only about one
to three hours.[24] The sperm from the early-spawning coral are
likely to be swept off the coral reef and be quite dilute at the
time the eggs from the late-spawning coral are released, thereby
reducing or preventing interbreeding between the two species.

Finally, an obvious yet perhaps underappreciated barrier to
species interbreeding is lack of attraction. We humans walk outside among squirrels, pigeons, grass, and trees, and yet we do
not attempt to breed with any of these other organisms. Certain
cues are required to cause an individual to elicit courtship behaviors or to accept courtship advances from another individual,
and even closely related species often differ in these cues. Such
cues can be transmitted from virtually any modality or combination of modalities: visual displays, sound, touch, etc. For
example, females of the closely related wolf spider species *Schizocosa ocreata* and *S. rovneri* recognize males as being of the same
species based on visual cues as well as vibratory cues transmitted
through tapping on the surface nearby.[25] Interestingly, researchers studied which visual cues were important in these species
by presenting spider females with videos of courting males and
observing the responses of the females.[26] Analogous to what
businesses do with human models in photo and television ad-
vertising, the researchers also manipulated the images in the
videos to make males more or less attractive to females of each
spider species.[27] Species-specific breeding-related cues may also
involve chemicals; for example, females of the closely related
fruit fly species *Drosophila simulans* and *D. sechellia* differ in their
pheromones, and *D. simulans* males are repulsed by the *D. sechellia* female pheromone. As such, *D. simulans* males will only
mate with females of their own species. However, females of the
two species can be experimentally crowded in a very small space

such that some of their pheromones rub onto each other.[28] Male *D. simulans* will then refuse to court *D. simulans* females "perfumed" by *D. sechellia*; in effect, they "smell wrong." In contrast, male *D. simulans* will sometimes court *D. sechellia* females that have been perfumed by *D. simulans* females.

Given that interspecies mating is uncommon on Earth, it appears unusually common among the humanoid species depicted in the various *Star Trek* series. Indeed, attraction to members of other humanoid species does not seem noticeably weaker in any of the five series than attraction to members of one's own species. Still, some instructive examples do appear. As noted in the *DS9* "Melora" example at the start of this section, doubts appear about the long-term possibilities of relationships with members of other species. Presumably, low-gravity-preferring Elaysians would rarely co-occur with Earth-gravity-preferring humans in the absence of mechanical intervention, suggesting that the two species would be spatially restricted and rarely interbreed. The same applies for species that breathe different gases. However, this episode identifies a challenge that practicing biologists face often: whether such spatial restriction (even associated with distinct adaptations or preferences) actively prevents interbreeding or whether species may, in the future, come into contact and potentially interbreed. The example in this episode appears to fall into the latter category, and cases like it pose a challenge for species definitions when no other barriers to reproduction exist.

Attraction via pheromones comes up in *Star Trek*, too. The pheromones of Orion females attract not just Orion males but also human males. However, the pheromone's effect was limited to only some humanoid species, since Vulcans appeared immune to its effects (*ENT* "Bound"). This pheromone effect on some species but not others is similar to what is observed with female *Drosophila sechellia*: its pheromones attract males of *D. sechellia* and related species *D. melanogaster*, but they fail to attract *D. simulans* males. A different chemical signal (not related to smell) in Elasian females also causes bonding by males and, again, this chemical seems to have an effect on humans (*TOS* "Elaan of Troyius"). However, despite the effect being permanent on Elasian males, human Captain Kirk was able to recover from his exposure to it, suggesting the effectiveness of this chemical signal may be weaker on males of other species.

Finally, cyclically timed mating bouts like those observed in corals and cicadas are also noted in *Star Trek*. Vulcans in the series undergo a neurological imbalance forcing them to engage in a ritualized mating every seven years (called *pon farr*, and first mentioned in *TOS* "Amok Time"). However, although this mating ritual has a periodicity, Vulcans in *Star Trek* mate outside of the pon farr as well, so this periodicity does not prevent mating with any other species. Indeed, some Vulcan-human hybrids are depicted in the series, most notably Science Officer Spock. The next section explores the fate of hybrids on Earth and in *Star Trek*.

FATE OF INTERSPECIES MATING

ENT, Season 4, Episode 20, "Demons"
ENT, Season 4, Episode 21, "Terra Prime"

A xenophobic, extremist group produce a hybrid daughter between a Vulcan and a human from their DNA via cloning. T'Pol, the Vulcan mother, learns that her daughter is dying. Doctor Phlox notes, "It's genetic. Her Vulcan and human DNA aren't compatible."

Despite the barriers to interbreeding discussed in the previous section, organisms do sometimes mate with other species. However, this mating does not always produce fertile adult offspring, meaning that transfer of genes from one species to another is not always possible. The poor fate of interspecies mating can occur at several stages, including failure of fertilization to occur or genetic incompatibilities in the hybrid leading to premature death or sterility, as proposed in the example from *ENT* "Terra Prime" above. I explore these fates in this section.

Sperm from animals or pollen from plants is often far more effective at fertilizing eggs/seeds of the same species than it is at fertilizing eggs/seeds from other species. Although humans tend to think of such fertilizations happening in the context of copulations, many species fertilize externally. As such, gametes are released into the air or water that fuse with other gametes of the same species. One of the best-studied cases of fertilization specificity comes from the gametes of two species of abalone (a marine shellfish). Fertilization in these species requires a sperm protein creating a hole in the egg surface protein so the sperm DNA can enter.[29] The red and pink abalone species differ in these proteins such that sperm from one is less able to create a hole in the egg surface of the other, thereby reducing the potential for interspecies fertilization.[30] Other times, barriers to fertilization may be separate from the gametes themselves but have to do with the physiology of the organism. For example, interspecies mating between some *Drosophila* fruit fly species results in an "insemination reaction" in the female: dramatic enlargement of the vagina and formation of an opaque structure inside it that prevents the female from being able to lay eggs.[31] A similar structure develops following within-species mating, but the structure is less dramatic and subsides within a few hours. In contrast, this structure lasts for several days after interspecies mating and essentially sterilizes the female during this period. This allergic-like reaction in females to semen from males of other species reduces the formation of hybrid offspring. Finally, mating between two wasp species (*Nasonia giraulti* and *N. longicornis*) exhibits a problem with similar outcome: offspring are not produced after interspecies mating. However, in this example the incompatibility is caused by differences in intracellular bacteria among the wasp species, and "curing" the wasps with antibiotics makes them fully compatible with the other species.[32]

Hybrids are sometimes formed successfully from interspecies mating, but genetic incompatibilities often still cause the hybrid to die at a young age and/or to be sterile. One interesting example of premature death associated with hybrids is that of the swordtail and platyfish (*Xiphophorus hellerii* and *X. maculatus*).[33] Platyfish possess black-pigmented spots that are controlled by two genes: one that creates the spot and one that prevents the spot from growing too large and becoming cancerous (a "tumor suppressor"). Swordtails lack the spots and have neither of these genes. If the species interbreed, the hybrid offspring inherit both the spot and the tumor suppressor variants, and thus have spots like the platyfish. However, if those hybrids breed with each other or with swordtails, some individuals inherit the spot gene variant but not the suppressor variant, and these individuals develop deadly tumors. As such, a subset of grandchildren of matings between swordfish and platyfish die prematurely. Despite this and many other examples, hybrids between some very distantly related species do occasionally survive to adulthood, with one of the most extreme examples being hybrids between two ferns that share an ancestor roughly 60 million years ago.[34]

Hybrids between species (especially distantly related ones) that survive to reproductive age are often sterile, with the most famous example (known for well over a century)[35] being the mule: the sterile hybrid offspring of a horse mother and donkey

Figure 10. A zonkey: the hybrid offspring of a zebra father and donkey mother. The author of this book thinks they look as though they are wearing socks. Photo by Ruth Louise Thompson, reproduced with permission, http://kayaksailor .deviantart.com/.

father. Many other sterile hybrids are known, such as the zonkey (cute hybrid offspring of a zebra father and donkey mother; see figure 10) or liger (giant hybrid offspring, sometimes weighing 400 kg, of a lion father and tiger mother). In some of these cases, the sterility is caused by the two species having different numbers of chromosomes (see chapter 3). However, sterility need not be complete in all of these hybrid offspring; female ligers are sometimes fertile, and even a handful of fertile female mules have been documented.[36]

In the examples above involving mammals, the female hybrid was the one that was more likely to be fertile. This same pattern is also striking among *Drosophila* fruit fly species—if one hybrid sex is fertile, it tends to be the females. However, birds and butterflies exhibit the opposite pattern: male hybrids are more

likely than female hybrids to be fertile. Sterility therefore cor-
responds to which sex carries two different "sex chromosomes."
Sex chromosomes—often called X and Y chromosomes—are
inherited differently in genetic males and females. In mammals
and fruit flies, offspring that inherit one X and one Y chromosome grow up as genetic males, while those that inherit two
X chromosomes grow up as genetic females. The opposite is
true in birds and butterflies: XX individuals are male and XY
individuals are female. Returning to the sterility of hybrids,
XX hybrids are far more likely to be fertile than XY hybrids—
among 223 species pairs for which one hybrid sex is fertile, the
XX hybrids are fertile in 213 pairs, and the XY hybrids are fertile in 10 pairs.[37] The same general trend comes up when examining which hybrid sex is more likely to survive to adulthood:
XX individuals are more likely to survive than XY individuals.
These patterns were first described in 1922,[38] and have since
been termed "Haldane's Rule." The same trend is even observed
among plants bearing sex chromosomes.[39] We now know that
sterility or inviability associated with Haldane's Rule seems to
have multiple genetic causes, most of which are associated with
having two different sex chromosomes.[40]

Barriers to successful fertilization between species are implied in the *Star Trek* series but not described in detail. For example, in *TNG* "The Emissary," one character notes regarding
the ability of humans and Klingons to have a child that "the
DNA is compatible, with a fair amount of help." While the comment is cryptic, DNA encodes proteins (discussed in chapter 3),
so perhaps there is a fertilization barrier between Klingons and
humans due to differences in sperm and egg proteins like that
observed between abalone species. The "help" may involve application of an additive that facilitates the sperm DNA enter-

ing the egg. Some unspecified interventions are also needed for successful Vulcan/human interbreeding as well (e.g., *ENT* "E²").

There is no direct mention of a potential incompatibility successfully preventing species interbreeding except in the case of the example mentioned at the start of this section (*ENT* "Terra Prime"). Interestingly, this incompatibility turns out not even to be a true species incompatibility but a flaw in the cloning process. The writers may have identified this cloning problem because of purported health issues associated with large mammal clones, such as the famous cloned sheep Dolly (who died prematurely in 2003, and the *ENT* episode aired in 2005), but other large mammal clones have not had such health issues.[41] More commonly in *Star Trek*, however, the characters are surprised at various unexpected biological compatibilities associated with intimacy among alien humanoids (e.g., *VOY* "The Disease," *ENT* "Unexpected").

Again, unusually many humanoid hybrids are observed in the various *Star Trek* series, affording us the opportunity to look for general patterns. Table 1 presents the hybrids that I was able to find that fit a few criteria.‡ To be included, the hybrid must have been successfully born (pregnancy alone was not counted) and must have resulted from the union of a sperm and egg (i.e., not been engineered from DNA or other cells like in the example at the start of this section). I focus on "first-generation" hybrids from two species: the immediate offspring of a mating between species 1 and species 2 rather than the later-generation descendants of those hybrids. For cases in which a series depicts

---

‡   I received a lot of help assembling this table, and many parts of this book, from entries in the Memory Alpha collaborative website: http://memory-alpha.wikia.com/.

Table 1: First-Generation Hybrids (of Known Gender) Depicted or Described in *Star Trek*

Species	Hybrid M/F	Ref
Human-Betazoid	F (<u>Deanna Troi</u>), F (Kestra Troi),	*TNG, TNG* "Dark Page"
	F (<u>mother of Devinoni Ral</u>)	*TNG* "The Price"
Human-Klingon	F (<u>B'Elanna Torres</u>), F (<u>K'Ehleyr</u>)	*VOY, TNG* "The Emissary"
Human-Ktarian	F (<u>Naomi Wildman</u>)	*VOY*
Human-Napean	M (Daniel Kwan)	*TNG* "Eye of the Beholder"
Human-Ocampan	F (<u>Linnis Paris</u>)	*VOY* "Before and After"
Human-Romulan	F (Sela), M (<u>father of Simon Tarses</u>)	*TNG, TNG* "Drumhead"
Human-Skagaran	F (<u>mother of Bethany</u>)	*ENT* "North Star"
Human-Vulcan	M (Spock), M (Lorian)	*TOS, ENT* "E^2"
Cardassian-Bajoran	F (Tora Ziyal), F (unnamed)	*DS9, DS9* "Covenant"
Cardassian-Kazon	M (Seska's baby)	*VOY* "Maneuvers"
Klingon-Romulan	F (Ba'el)	*TNG* "Birthright"
Survival overall: 12 F, 5 M.	Survival by category: 7 F, 3 M, 1 both.	
Fertility overall: 7 F, 1 M.	Fertility by category: 5 F, 1 M.	

Note: F, female; M, male; both, at least one male and at least one female observed. Underlined hybrids produced offspring in at least one timeline depicted in the series.

later-generation hybrids, I list the pairing only if we can infer the sex of the first-generation hybrid ancestor.

Looking at the suite of hybrids listed in the table, we see more female hybrids than male. Assuming males and females are produced at similar frequency at the time of conception, then the difference in appearance (while not dramatic) could illustrate Haldane's Rule for survival: female hybrids may be more likely to survive than male hybrids. Sterility comes up a few times in

Star Trek (*TOS* "Wink of an Eye," *TNG* "When the Bough Breaks"), but not in the context of interspecies hybrids. However, the table also highlights the hybrids known to produce offspring (demonstrating fertility) in the series. Here, the pattern is more striking: all but one of the first-generation hybrids depicted or described in the series who produced offspring were female. If we speculate that this depiction reflects a difference in hybrid fertility, meaning at least some of the hybrid males were sterile, then we may be observing some signal of Haldane's Rule just like among species on Earth. I do not think that the *Star Trek* writers did this intentionally, but the coincidence is amusing.

DETECTING DESCENDANTS OF INTRA- AND INTERSPECIES MATINGS

Traditionally, interspecies matings were identified by observing them directly or inferred by detecting intermediate forms in the wild. However, the former approach is inefficient, and the latter may be complicated by variation within species; for example, the author of this book has more body hair than many humans, but he is not a human-chimpanzee hybrid (and has genetic data to prove it).

Over the past 50 years, starting around the time *TOS* aired, scientists have used genetic data increasingly for making evolutionary inferences. More specifically, "DNA markers" became ever more popular with evolutionary biologists in the 1980s and 1990s (around the time *TNG* launched) to study many questions, including paternity, in wild animal populations. One of the early insights from these studies was that many species that appear socially monogamous—one male and one female bond and spend most of their time together—often have frequent "extra-pair copulations" (breeding with individuals besides their partners). Among birds, for example, early work suggested that over 90% of bird species were monogamous, but later analyses suggested that "extra-pair" offspring are found in 90% of bird species, with over 10% of offspring from supposedly monogamous species being from extra-pair copulations.[42]

Genetic data over these past few decades have also showed that many species are slightly "leaky."[43] Although most individuals breed with others of the same species, a subset of individuals breeds with other species and thus introduces foreign genes into the species. Genetic signatures of such interbreeding have been detected even among ancient Earth hominid species. Whole-genome sequences are available from many humans,[44] as well as a few fossil Neanderthals and Denisovans (extinct hominids: see figure 5 in chapter 2),[45] and statistical comparisons of these sequences strongly suggest that modern humans have some Neanderthal and Denisovan ancestry, likely dating to roughly 50,000 years ago.[46] Modern humans may have as much as 3% of their genetic material from ancient Neanderthals and/or 5% of their genetic material from ancient Denisovans,[47] and commercial companies such as 23andMe will even provide customers with an estimate of their Neanderthal genome fraction. Genetic exchange between these species was not always free from consequence to humans—some evidence suggests that many genes coming from Neanderthals and Denisovans were at least slightly detrimental to humans, perhaps through reducing male fertility,[48] or introduction of various genetic diseases and disease predispositions.[49] However, some adaptive gene variants were also introduced to humans from Neanderthals or Denisovans, with an example being one of the gene variants mentioned in chapter 4 conferring the ability to get enough oxygen at high altitudes in modern ethnic Tibetans.[50]

The later *Star Trek* series (after *TOS*) leveraged genetic approaches to infer parentage as well as mixed-species ancestry. For example, Doctor Crusher used a DNA test to determine if Jason Vigo was the son of Captain Picard in *TNG* "Bloodlines." Doctor Phlox in *ENT* used genetic tests often, such as to determine that the hair follicle obtained in "Demons" came from the daughter of Science Officer T'Pol and Chief Engineer Trip Tucker. Doctor Phlox also used genetic markers to identify Karyn Archer as the great-granddaughter of Captain Jonathan Archer and to determine that Lorian was the son of T'Pol and Trip Tucker in "E²." This episode also had several tests for species ancestry, and Doctor Phlox noted that Karyn Archer possessed DNA from three non-human species (presumably not including trace amounts of Neanderthal or Denisovan). Genetic exchange between humanoid species apparently becomes common in the *Star Trek*–envisioned future, as illustrated by thirty-first-century operative Crewman Daniels referring to himself as "more or less" human (*ENT* "Cold Front").

The depiction of genetic exchange among humanoid life in the coming millennium in *Star Trek* is thus similar to what really happened with humanoids on Earth roughly 50,000 years ago. While much of this similarity is probably coincidental, I am impressed with the writers' imagination and their accuracy in depicting how genetic tests are performed to test for parentage or genetic exchange among species. Sometimes how scientists approach their trade is bounded by their imagination, and science fiction can both energize and foster thinking "outside the box" in ways that lead to progress. I explore this briefly in the next (final) chapter of this book.

CHAPTER 6

SCIENCE VERSUS SCIENCE FICTION

As a practicing scientist, I have largely assumed that principles observed in life on Earth should apply (to some extent) to life that may have originated and evolved extraterrestrially. This assumption has both strengths and weaknesses. On the positive side, some principles derived from the study of life on Earth must be applicable more broadly to life that may have evolved on extraterrestrial worlds. Scientific investigation would be extremely difficult if every observation of a pattern was a "special case," and the ability to extrapolate from limited data is a time-tested foundation of science. Just as physics presumably works the same on Earth as it does on Mars, surely at least some principles of evolution apply to extraterrestrial life just as they do life on Earth. As discussed in chapter 4, natural selection is a mathematical inevitability if there is inherited variation that affects the number or survival of offspring.

However, other aspects of life on Earth covered in this book may not apply elsewhere. Life forms may have inheritance or physiology completely distinct from what we have seen on Earth. For example, "memories" may well be encoded in their inherited code (their equivalent of genes), and they may possess the means to exchange or combine their code in manners completely unlike the sexual reproduction we observe on Earth. Again, we have only observed life forms descended from a single origin, and truly independent life forms may well follow largely different rules.

Therein lies the distinction between science and science fiction more generally, as well as the relative strength of each. Science often assumes that documented principles are constant and general, at least until proven otherwise. In contrast, science fiction captures the imagination and encourages thinking of these principles as starting points but potentially incomplete. In essence, science fiction is an exercise in "what ifs" that question presumed limitations as we know them today. The *Star Trek* series attempt to explain their imagined scenarios with potentially realistic scientific explanations—their casts are curious explorers not just in charting new planets but also in researching and trying to understand the world around them better through science. In contrast, many other series (and occasionally *Star Trek*, too) also depict fantastical events but ignore or dismiss deviations from known scientific principles (or fall back on an equivalent of "magic").

While science obviously influences science fiction, science fiction also has elements to offer science. First, along the lines of broad thinking, science fiction helps people appreciate scientific curiosity and the importance of basic research in general. Second, science fiction, and *Star Trek* in particular, has inspired generations of people to engage with science or even become practicing scientists. I will discuss each of these in turn.

"BASIC" RESEARCH

TOS, Season 2, Episode 19, "The Immunity Syndrome"
The *USS Enterprise* encounters an enormous space-faring, single-celled organism that appears to have killed the crew

of another ship as well as all the inhabitants of a solar system. The crew are successful in destroying the organism and escaping safely. Mr. Spock notes that he has "some fascinating data on the organism." Doctor McCoy chides him because he botched one of the tests he performed on the organism.

The episode described above was one of the few in *TOS* where distinctly different "life" was encountered by the crew. The initial reason for studying this new life was not curiosity. The organism posed a threat to the ship and to other solar systems, and the crew sought to mitigate this threat. In the process, they engaged in scientific thinking—examining data, devising predictions and tests, and then executing the tests and re-evaluating. However, until the last few minutes, most of the exploration was "application oriented": kill the life form before it kills them and others. This process is not unlike a physician seeking to determine the cause of an ailment and curing the patient, and the episode's script drew a similar analogy.

However, something different happened at the very end. The organism was destroyed. The threat was neutralized. This appeared to have been a unique life form, so no future application seemed apparent. When physicians heal patients, they do not typically dig further into the nature of the past ailments. Instead, Mr. Spock appeared eager to examine the data obtained about the organism, and Doctor McCoy noted a desire for more data yet. Why?

"Basic research" is the expression used to describe scientific studies that seek to improve our understanding of nature. Such research may focus on garnering evidence for or against existing theories, or it may pose altogether new questions and suggest answers. The immediate goal of such research is often not to solve an existing problem affecting people—not to cure a disease or improve crop yields, for example. Instead, such research is often motivated by curiosity to improve our knowledge of the world around us, ideally elucidating grand principles common to much of life (e.g., natural selection) or identifying exceptions to (or flaws in) existing principles. The implication in the episode above was that Mr. Spock and Doctor McCoy anticipated studying the data obtained in the context of basic research. Such an implication is highly consistent with the stated mission of the crew as mentioned in chapter 1: "to seek out new life,"* not "to seek out life that is identifiably useful to us a priori."

Legislators sometimes question whether we should allocate limited financial resources to basic research and exploration when nations already have clear needs that can be helped with application-oriented research (e.g., disease control, food security). In 2008, United States vice presidential candidate Sarah Palin mocked tax dollars funding "fruit fly research in Paris, France."[1] The contributions of fruit fly research to medicine and other areas have been enormous—leading to six Nobel prizes as of 2017!—and could fill a book of their own, but I would like to give one lesser-known example that I find particularly striking.

In the 1980s and 1990s, teams of researchers discovered fruit flies infected with bacteria called *Wolbachia*.[2] Their interest was not in its clinical relevance, but in an interesting evolutionary dynamic that resulted from infection. *Wolbachia* is transmitted in fruit flies from mother to offspring. Female fruit flies that are infected with *Wolbachia* can produce offspring with either infected or uninfected males. However, female fruit flies lacking infection cannot produce offspring from males bearing the infection. As such, an evolutionary advantage exists for females

* From introductory sequence of each episode of *TOS* and *TNG*.

to be infected (those females can always make offspring) and, as a result, infections spread rapidly across natural populations. Building on this work, other researchers found that a strain of *Wolbachia* from fruit flies can be artificially introduced into mosquitoes and spread, due to a similar evolutionary dynamic, across populations.[3] More strikingly, infection by *Wolbachia* reduces transmission of various viruses by the mosquitoes, including dengue, chikungunya,[4] and Zika.[5] Since infection by mosquito-transmitted dengue alone poses a threat to over 2 billion people on our planet (infecting 50 million people annually),[6] control strategies are vitally needed. As such, researchers are examining the relative risks of releasing *Wolbachia*-infected mosquitoes to locally eradicate these diseases,[7] and are already beginning some controlled local releases to that end.[8] In the long run, basic research on a curious evolutionary dynamic in natural populations of a fruit fly may well lead to disease control strategies that will save millions of human lives.

Other studies of nature have already had a dramatic impact on human health and many human activities. In the mid-1960s, a professor and his undergraduate were examining energy production (photosynthesis; see chapter 1) in bacteria that lived in hot springs at Yellowstone National Park in the USA. They described a new species called *Thermus aquaticus*, which was able to grow at the unusually high temperature of 79°C.[9] The student went on to show that enzymes from this species tolerate even higher temperatures.[10] The stability of an enzyme from *Thermus aquaticus* (called "Taq polymerase") became the key ingredient, 15 years later, in a procedure called the "polymerase chain reaction" (PCR).[11] For the past 30 years, PCR has been one of the most widely used molecular biology approaches for amplifying a segment of a genome from a very small DNA sample. This procedure is vital to virtually any scientific investigation, and the discovery of PCR caused a renaissance for molecular biology and any fields leveraging it. PCR has been used heavily in the discovery of and screens for disease-causing genes,[12] food science, criminal forensic investigations, genetic engineering, and many, many other applications. Simply put, most progress in genetics and molecular biology that enriches our lives and livelihoods would have been set back a decade or longer without PCR. And PCR would not have been developed without basic research looking at what bacterial species may be lurking in hot springs at Yellowstone National Park.

Many other world-changing discoveries would not have been possible without the pursuit of curiosity and discovery in science: X-rays, penicillin, the vaccine for polio, and more.[13] Both private and public funding agencies recognize the importance of basic exploratory research to addressing basic human needs such as food security and health, and even targeted agencies such as the National Institutes of Health advocate for and fund basic scientific research in addition to application-based research.[14] They know that basic research provides the tools for application-based research to solve the problems of the world, and many more such tools have yet to be discovered from this kind of research. Economic studies, too, have shown that public (i.e., national via taxpayers) investment in basic research leads to a positive return on investment, though the rate of return is difficult to quantify since the return can take many forms (e.g., training skilled graduates, creating new companies and industries).[15]

Like these agencies, the various *Star Trek* series also strongly advocate for basic research and exploration, even beyond the core mission described in the introductory sequence. Science Officer Spock in *TOS* is always keen to explore new life, sometimes even putting himself or the crew at risk to do so (e.g., *TOS* "Devil in the Dark"), and the crew sometimes mock how often he describes

observations as "fascinating" (*TOS* "The Ultimate Computer").
The crew of the *USS Voyager* repeatedly express excitement at the
prospect of encountering new species (e.g., *VOY* "Innocence").
When one crew member criticizes Captain Janeway for wasting
too much time exploring rather than focusing on getting the
crew home, she counters, "We seek out new races because we
want to, not because we're following protocols. We have an insa-
tiable curiosity about the universe" (*VOY* "Random Thoughts").
Finally, outside-the-box thinking is repeatedly encouraged. In
the series finale for *TNG*, Q tells Captain Picard, "For that one
fraction of a second, you were open to options you had never
considered. That is the exploration that awaits you. Not mapping
stars and studying nebulae, but charting the unknowable possibilities of existence" (*TNG* "All Good Things . . .").

I argue that the promotion of basic research in the public
eye (as done in *Star Trek* and some other science fiction) advances science by raising awareness that basic science is generally important to a society and thus merits support, and by
inspiring people to consider studying (and perhaps specializing
in) science by countering misconceptions that it is boring or
too difficult. The first of these points is vital to the continua-
tion of science since public funds—i.e., government support via
taxes—support much basic research. I explore the latter in the
final section below.

INSPIRATIONS BY AND DEDICATIONS TO *STAR TREK* IN BIOLOGY

Although many students avoid "STEM" (science, technology, engineering, mathematics) disciplines because of perceptions that
the subjects are boring or difficult,[16] several people who came
to work in these areas cite *Star Trek* as significant in their final

directions. One excellent example is Doctor Mae Jemison, the first African American woman to travel in space. Doctor Jemison noted in an interview that *TOS* and *TNG* gave a science-minded generation an outlet for expression,[17] and she herself went on to appear in an episode of *TNG* ("Second Chances"). Other astronaut-fans eagerly appeared in the *Star Trek* series as well, including Colonel E. Michael Fincke and Colonel Terry Virts, Jr. (*ENT* "These Are the Voyages . . .").

While we may anticipate a link between science fiction series set in space and becoming an astronaut, the influence of science fiction in general and *Star Trek* in particular is much broader across STEM disciplines. Physicist Doctor Stephen Hawking once remarked in an interview that "science fiction is useful both for stimulating the imagination and for defusing fear of the future."[18] Following on this sentiment, Sigma Xi, the scientific research society, asked its broad STEM membership if and how they were influenced by science fiction. The responses were overwhelmingly positive, and many of the replies cited influence by one or more of the *Star Trek* series. Respondents ranged in age and in profession: chemists, engineers, psychologists, geologists, and biologists. One respondent who worked at the US Centers for Disease Control and Prevention wrote, "truthfully *Star Trek* was the biggest motivator! I was enthralled by [*TOS*] Doctor McCoy's medical instrumentation, Spock's logic and scientific outlook and the psychology involved in dealing with people from other cultures. Appreciating IDIC—Infinite Diversity in Infinite Combinations—has served me well in my scientific career."[19]

Some modern pieces of technology resemble items used in *Star Trek* series filmed much earlier: flip-phones from the early 2000s resemble the *TOS* communicators depicted in the 1960s, tablets used today resemble the "PADDs" used in *TNG* and the later series beginning in the 1980s and 1990s, etc. This resemblance could suggest that the inventors of the modern pieces were consciously or unconsciously influenced by *Star Trek* and other science fiction. Lesser known, however, is that many biomedical devices and approaches we possess today also mirror those in *Star Trek* and other science fiction, suggesting an influence. One study analyzed more than 50 science fiction movies (including several *Star Trek* movies) and found "a definite non-linear correlation between the bioinstrumentation shown in science fiction films and the development of biomedical instrumentation."[20] Recently, the XPrize and Qualcomm foundations awarded millions of dollars to research teams developing a "tricorder" (a portable health diagnostic system) similar to those depicted in *Star Trek*. Even the name was retained, making the inspiration obvious.

Given the love of *Star Trek* by scientist-fans, it is unsurprising that many have dedicated elements of their work to characters in the series. Some examples include:

- A wasp whose individuals vary greatly in appearance named *Phanuromyia odo*, inspired by shape-changing character Odo in *DS9*;[21]
- A beetle named *Agra dax* after the character Jadzia Dax in *DS9* and dedicated to the actress who played her;[22]
- A hermit crab with a wrinkled belly area named *Annutidiogenes worfi* after the *TNG/DS9* character Worf, who bore a wrinkled forehead;[23]
- A clam named *Ledella spocki* named after *TOS* character Spock and having valves shaped like the pointed ears that Spock possessed in the series;[24]
- Two DNA sequences that have the ability to copy themselves to other locations in the genome named Worf and Spock;[25]
- A popular software package used for analysis of DNA sequences named Picard, after the captain in *TNG*.[26]

The excitement and inspiration of these scientists by science fiction can be further passed on to new generations. Science fiction conventions increasingly have "science tracks," wherein practicing scientists present "the science behind" their favorite movies and television series. The science fiction and popular culture convention "DragonCon," held annually in Atlanta, Georgia, USA, hosts an active science track, with panel titles for 2017 ranging from "*Star Wars* Biology vs. *Star Trek* Biology" and "Chemical Warfare in Wonder Woman" to "Ancient Viruses in the Permafrost" and "Personal Genetic Testing." More than 80,000 people attended this convention. Academic scientists also use science fiction to aid in teaching fundamental concepts in their classrooms. The author of this book cotaught a class at Duke University called "The Biology of Popular Science Fiction and Movies," and others have run similar courses elsewhere.[27] The principle is to leverage an engaging medium in which students are already interested to explore concepts that, by themselves, may sometimes be perceived by students a priori as abstract or dry. This book is an example of such an effort to engage a broader audience in science, and I sincerely hope that is has been successful in engaging the readers in the exciting evolutionary and genetic science underlying portrayals in *Star Trek*.

EPILOGUE

Star Trek is a science fiction franchise. Good science fiction stimulates the curiosity by posing hypotheses that extend beyond known truths. I have argued in this chapter that the various *Star Trek* series depict the sort of scientific curiosity that is typical in basic scientific research, helping viewers appreciate the excitement associated with and value of such approaches. I have also argued that science fiction in general and *Star Trek* in particular have influenced a subset of people to explore science further, seeking to identify further truths. I count myself among these people so influenced and, again, I hope this book serves as a vehicle by which others get similarly excited while also learning a lot about basic concepts and recent findings. While my focus in this book has been on presenting concepts from evolutionary biology (with a healthy amount of genetics mixed in), the same has been or could be explored of *Star Trek* with respect to ethics or physics or engineering, or many other disciplines. We are all students of truth, and our pursuits and imaginations are often stimulated by the scientifically grounded creativity associated with *Star Trek* and other good science fiction. I close this chapter with an amusing quote from Garak in *DS9* "Improbable Cause" that perhaps justifies some of the scientific leaps taken in the series: "The truth is usually just an excuse for a lack of imagination."

APPENDIX

MINING GEMS AND COAL

The chapters of this book present a (hopefully) coherent narrative for learning some basic evolutionary biology using *Star Trek* episodes and movies. This appendix is devoted to a few topical scientific observations from *Star Trek* episodes relevant to the areas discussed in the book, in no particular order. In some cases, *Star Trek* did a particularly good job in its portrayal of the science (the gems), but in other cases, high-school-level knowledge of biology could have prevented a misportrayal (coal). A few cases of each are mentioned here, at varying depth. I present these as fun examples rather than as a comprehensive assessment of evolutionary or genetic mentions in *Star Trek*.

COAL: PLANTS ARE NOT ALIVE?

TOS, Season 2, Episode 22, "By Any Other Name"

The *USS Enterprise* is orbiting a planet with large green continents and blue bodies of water. As the landing party transports down, they see trees, bushes, and grass all around. Science Officer Spock examines his tricorder scanner and immediately reports, "No life form readings, Captain."

This happened more than once in *TOS* (see also *TOS* "The Alternative Factor," "Who Mourns For Adonais?"): plants were not regarded as "life forms" when these scans were conducted. In some cases, the crew went back and forth between saying "animal life" and just "life" as though they are synonymous (*TOS* "Shore Leave"). Generally speaking, the crew was unusually incurious at the sheer number of planets bearing plant life. How did the plants get there? Did they evolve in situ or were they introduced to so many planets? In what ways are they similar to or different from plant life on Earth? So many questions could have been asked but never were. Fortunately, there was greater emphasis on the full spectrum of life in later iterations—scientist Carol Marcus was emphatic that even "a microbe" counted as life when she had a team investigating planets (*Star Trek II: The Wrath of Khan*).

GEM: (TRILL) SYMBIOSES

TNG, Season 4, Episode 23, "The Host"
DS9, Season 2, Episode 4, "Invasive Procedures"
DS9, Season 3, Episode 4, "Equilibrium"

The Trill in *Star Trek* are a humanoid species from the planet of the same name. A small fraction of these humanoids are "joined" with large (rat-sized) slug-shaped symbionts that live within their abdomens. Within a few days, the organisms become dependent on each other, such that the symbiont cannot be removed unless put into another host, and the host can no longer live without the symbiont. The two organisms further share a single consciousness and memories.

While direct sharing of memories is not observed in life on Earth (as far as we know), symbioses are common, and I applaud *Star Trek* for highlighting a fictional example. Focusing on humans in particular, we have a diverse array of microbes in different parts of our bodies, prompting some scientists to describe the parts of the human body as a series of "ecosystems" for such microbes. The number of such cells in and on the human body is hard to imagine, reaching 100 trillion microbes in one individual's intestines alone.[1] Many of these microbial symbionts appear to affect digestion and nutrient absorption, immune response, vitamin synthesis, and likelihood of contracting various diseases.[2] Healthy individual humans differ remarkably in their microbiomes even when comparing the same parts of their bodies,[3] and the composition is partially influenced by diet, activity, and other environmental effects.[4] In effect, we humans are "joined species" with these many symbionts, and harm to them can cause harm to us. One of the simplest observations of this fact is that women often get yeast infections after taking broad-spectrum oral antibiotics that kill some of the normal bacterial flora.[5]

The humanoid Trill in *Star Trek* are called a "joined species" with their slug-shaped symbionts, and some examples of such combined organisms are known on Earth. The most famous is arguably the lichen: a combination of fungi providing structure with algae or blue-green algae providing energy via photosynthesis. Like humans, lichens also harbor diverse bacteria and other microbes, which may work with the elements mentioned above.[6] Living as a lichen symbiont appears to be a successful way for fungi to acquire nutrients given that roughly one in five fungal species has evolved such associations.[7] In *Star Trek*, not all humanoid Trill are associated with symbionts either: Lieutenant Dax said, "Only one Trill in ten is chosen to be joined," and notes that many who do not undergo symbiosis lead fulfilling lives (*DS9* "Invasive Procedures").* Clearly, the Trill symbiosis is not

---

* Though the fraction is assumed to be much lower in *DS9* "Equilibrium."

exactly like anything observed on Earth but, again, I applaud the *Star Trek* authors for introducing the concept to a broader audience.

MIX: FUSING DIFFERENT SPECIES INTO ONE

VOY, Season 2, Episode 24, "Tuvix"

Lieutenant Tuvok (a Vulcan) and Neelix (a Talaxian) are sampling flowers from a planet's surface. When *USS Voyager* uses the transporter to return them back to the ship, the two are mysteriously merged at the molecular level into one healthy person, subsequently named Tuvix. Tuvix suggests this could be a case of "symbiogenesis." Captain Janeway explains, "Symbiogenesis is a rare reproductive process. Instead of pollination or mating, symbiogenetic organisms merge with a second species." Tuvix elaborates, "They're able to merge with other single-celled organisms to form a third unique species, a hybrid."

Biologist Doctor Lynn Margulis defined "symbiogenesis" as instances in which long-term associations between members of different species give rise to new behaviors, new tissues, or new species.[8] However, this definition is arguably broader than the one most biologists would use, which tends to focus on the formation of new taxa (such as lichens, described in the preceding section) or the endosymbiotic origin of mitochondria (see chapter 2) or chloroplasts. On the positive side, there are good examples of symbiogenesis on Earth, and fusing Vulcan and Talaxian genes into one individual of a new species could be considered such an event. On the negative side, Captain Janeway's description of it as a "reproductive process" or an alternative to mating or pollination is imprecise—while the process yields new taxa, it fails to perpetuate the parental species and therefore cannot be sustained as a reproductive process per se.

Setting aside transporter accidents, interspecies mating can result in merging genomes with a second species to create a new species. Many plant species in particular experience rare instances of "allopolyploidy": formation of a new species through the breeding of two different species wherein the resulting new species has three or more copies of its genes. Imagine species A has two copies of its genes, and species B has two copies of its genes. Rather than forming sperm and egg (or pollen and seed) with just one copy in each, the gametes retain both copies, and the resultant offspring is a new species that has four gene copies: two derived from species A and two derived from species B. This new species then reproduces either asexually or sexually with other polyploids derived from the same two parent species. If sexually reproducing, the new polyploid often cannot form fertile hybrids when breeding with the parent types, so it is truly a new species distinct from either parent. Some species with which we are very familiar formed via allopolyploidy, such as soybean and cotton.[9]

Intriguingly, allopolyploid hybrids between species are more likely to be fertile than regular hybrids between those same species. When regular hybrids form gametes, the chromosomes from the two parent species are sometimes too different to pair correctly, resulting in decreased fertility. However, polyploid hybrids have two copies of each chromosome from each parental species, allowing for better pairing of chromosomes.[10] Hence, fertile allopolyploid hybrids are sometimes formed between more distantly related species than fertile regular hybrids.[11] Like an allopolyploid, Tuvix presumably had both copies of Lieutenant Tuvok's and Neelix's genes (rather than just one copy of each, as if he had been their offspring; see chapter 3). Although we do not know whether Tuvix was fertile, carrying this double genome may be consistent with his "surprisingly healthy" status.

Despite my comparison to allopolyploids above, Tuvix was not formed by a mating between a Vulcan and a Talaxian, so the analogy is imperfect. Further, the explanation for why he formed (the plant transported with them had "lysosomal enzymes") was rather weak. Lysosomal enzymes function in digestion and other basic biological processes, and such enzymes are already present in all of the humans that have transported in this manner. The cause of this fusion is unclear at best.

GEM: PARASITOID REPRODUCTION

ENT, Season 1, Episode 4, "Unexpected"

Chief Engineer Trip Tucker visits a Xyrillian ship for a few days to assist their crew with repairs. His hands come into indirect physical contact with a Xyrillian female's hands while on the ship. Mr. Tucker later discovers strange growths on his body, and that the contact with the female implanted a Xyrillian embryo within him. Doctor Phlox notes, "It's not technically your child. . . . When reproducing, the Xyrillians only utilize the genetic material of the mother. The males simply serve as hosts." Mr. Tucker subsequently exhibits protective behaviors regarding the safety of potential children on the ship, even though the ship has no children on board.

In this fictional example, the child only carries genes from the mother's species, so Mr. Tucker does not benefit genetically from the reproduction. Mr. Tucker also lacks any nontechnological means for giving birth to a child, meaning that the child's emergence without technological assistance may be harmful or lethal to him as a host. These facets are typical for "parasitoid" reproduction on Earth: where an organism has its young develop inside another organism (the "host"), often killing the host upon emergence. Many wasp species are parasitoids wherein the females lay their eggs on other unsuspecting insects (e.g., caterpillars, beetles, fly larvae). The wasp larvae consume the host from the inside and eventually emerge after they have developed. Like the case with Mr. Tucker, infection with a parasitoid sometimes alters the host's behavior; for example, ladybugs infected with a larva from the green-eyed wasp *Dinocampus coccinellae* are temporarily paralyzed and involuntarily twitch to ward off predators until the fully developed wasp emerges, thereby serving as "zombie babysitters."[12] We do not know what would have happened to Mr. Tucker if the child was not removed from his abdomen with technology—perhaps he, too, would have become a zombie babysitter. But the writers are again applauded for their depiction of less-familiar forms of reproduction that are also observed in species here on Earth.

MIX: SPACE FUNGI AND "SPORE DRIVE"

(This example connects to the next one.)

DIS, Season 1, Episode 3, "Context Is for Kings"

A *USS Discovery* shuttle approaches a damaged starship (the *USS Glenn*). Astromycologist Lieutenant Paul Stamets notes that there is evidence of a "catastrophic basidiosac rupture" on the ship's hull. Michael Burnham, seeking to understand the discussion, noted that basidia are structures that produce spores. Lieutenant Stamets mentions that spores are the progenitors of panspermia. Later in the episode, Burnham learns that both the *USS Glenn* and

USS Discovery were developing an experimental "spore drive" organic propulsion system for virtually instantaneous travel. These spores come from the fungus *Prototaxites stellaviatori*. Apparently the ships use spores to travel on a "mycelial network" extending across the entire cosmos, and travel along this network results in the ship's walls getting damp.

This idea is . . . creative? Let me start with a few positive elements. Fungi do often reproduce via spores. These spores germinate to produce offspring which then grow and spread in part via masses of threadlike structures called mycelia. Basidia are indeed structures that produce the spores on some kinds of fungi (e.g., mushrooms), and in *DIS* "Choose Your Pain", *Prototaxites stellaviatori* is called a mushroom. *Prototaxites* was an ancient (now extinct on Earth) genus—it was the largest organism present on land for a long period around 400 million years ago.[13] And as discussed in chapter 2, some kinds of spores are often quite resilient to harsh conditions. Spores may have been associated with panspermia (see chapter 2). So, some of the terminology is used appropriately here, and the ideas are based on real science.

The mention of a "mycelial network" implies that these spores are connected in some way (perhaps by sheer density or by actual threads). Building on the episode described above, Lieutenant Paul Stamets (who, incidentally, is named after a present-day, real-world fungal biologist) notes that this mycelial network is a "discrete subspace domain" that spreads across the universe "fanning out into infinity" in *DIS* "Choose Your Pain." This subspace domain is then used as a basis for transportation—specifically, a giant tardigrade-like beast (called Ripper: more on it below) has a symbiotic relationship with the mycelia that allows it to navigate this network with instantaneous (quantum?) transportation. Um, okay.

The writers took clear liberties with the biology (and surely the physics) here. How did these fungal spores spread throughout the universe? Fungi evolved on Earth (see arguments in chapter 2 for humanoids—the same apply to fungi), and they are products of the same endosymbiosis event that spawned plants and animals (see chapter 2). Fungal spores may have spread to other worlds and thus contributed to panspermia, but they would have had ancestors on Earth and spread to other worlds later. Further, even though we know ancient *Prototaxites* existed on Earth, how did its spores become so abundant across the universe? One can reasonably suggest that such spores may have spread via meteorite impacts, but not so many as to cover the known universe (subspace or not) at any nontrivially low density. Even if we were to assume these spores did spread, they would certainly completely dry out in space. Why then does use of the spore drive result in damp walls?

Since I am already beating a dead horse, I will add briefly that we do not yet fully understand what *Prototaxites* actually was on Earth—it may not be from the group of fungi that even had basidia.[14] Nonetheless, parts of this presentation were based on real-world biology, albeit with some (extreme) creative liberties in terms of the space and instantaneous transportation elements.

MIX: HORIZONTAL GENE TRANSFER AND SPACE TARDIGRADE

(This example builds upon the one above.)

> *DIS*, Season 1, Episode 4, "The Butcher's Knife Cares Not for the Lamb's Cry"
> *DIS*, Season 1, Episode 5, "Choose Your Pain"

The *USS Discovery* crew captures a very large multilegged, barrel-bodied beast from the damaged *USS Glenn* in the episode discussed previously, and they name it "Ripper."

Michael Burnham examines Ripper and notes, "Your Ripper appears to share some natural traits with the tardigrade species, a docile creature that lives in the waters of the Earth. A micro-animal capable of surviving extreme heat and sub-freezing temperatures." Lieutenant Stamets notes that the beast can hold the coordinates of every charted star system in its head. In the latter episode, the crew explore further how the tardigrade leverages the spores to control travel. Michael Burnham explains, "Like its microscopic cousins on Earth, the tardigrade is able to incorporate foreign DNA into its own genome via horizontal gene transfer," and that this transfer helps it travel the mycelial network.

Like the previous one, many elements of this episode leverage published biology. I will focus on the "horizontal gene transfer" aspect first. While individuals of most species inherit most of their DNA from ancestors, individuals can occasionally incorporate segments of DNA from other individuals or even other species. Bacteria are well known to take up and incorporate DNA from their environment—indeed, studying this uptake is how DNA was first discovered to be the hereditary material of life.[15] However, DNA is also occasionally transferred into the genomes of animals, plants, or fungi, particularly if the species of origin lives in close association with a host species (e.g., from bacteria or viruses into animal hosts). While most of these transfers involve one or a few genes, they can sometimes involve much larger segments; for example, the full genome of a parasitic bacterium (*Wolbachia*—see chapter 5) was transferred into the genome one of its fruit fly species hosts.[16] Some horizontal transfers have been important to the success of the recipient individual or species. For example, a gene causing red carotenoid pigment production transferred from a fungus into the germ line of an aphid species,[17] and this gene now affects these aphids' susceptibility to predators. More negatively, some researchers have inferred DNA transfers from bacteria associated with cancerous human tissues.[18] Nonetheless, despite the evolutionary importance of some individual cases, horizontal gene transfers must not be very common into the germ lines animals, plants, or fungi. We can deduce this rarity because, in thousands of studies of thousands of different species, DNA-based phylogenetic trees from different genes typically estimate largely the same relationships as each other, and largely the same relationships as ones inferred from overall physical similarity (see chapter 2).

Enter the tardigrade (see figure 11 and also discussion of tardigrades in chapter 1). In 2015, a study was published suggesting that one in every six genes in a tardigrade species entered via horizontal gene transfer. Based on this inference, the authors further speculated that "animals that can survive extremes may be particularly prone to acquiring foreign genes."[19] As discussed in chapter 3 in the context of radiation, extremely stressful conditions may cause damage to an organism's DNA. Hence, the authors inferred that such damage may make species surviving extreme conditions more likely to incorporate foreign DNA. Publicity around this study likely influenced the writing of these *DIS* episodes. Unfortunately, subsequent research the next year failed to support the original findings. These authors inferred that horizontal gene transfer (HGT) "accounts for at most 1–2% of genes and that the proposal that one-sixth of tardigrade genes originate from functional HGT events is an artifact of undetected contamination."[20] The original tardigrade study was not unique in this regard—many other purported cases of horizontal gene transfer have been questioned.[21] Overall, while we know HGT occurs rarely and may have evolutionary impacts, tardigrades do not appear to have unusually high rates of it.

Figure 11. The tardigrade *Milnesium tardigradum*. Reproduced under the Creative Commons Attribution (CC BY) license from E. Schokraie, U. Warnken, A. Hotz-wagenblatt, et al. (2012) Comparative proteome analysis of *Milnesium tardigradum* in early embryonic state versus adults in active and anhydrobiotic state. *PLoS ONE* 7(9): e45682. doi:10.1371/journal.pone.0045682.

Let me briefly explore the "tardigrade" Ripper further. Initially, Michael Burnham did not say that Ripper is a tardigrade or that it is even related to one, but she said that it "appears to share some natural traits with the tardigrade species." However, subsequent references to the beast were often simply "the tardigrade." Ripper's similarity to tardigrade species on Earth mostly relates to external body shape and ability to tolerate extreme, space-like conditions. First and foremost, Ripper is a vertebrate—we see the vertebrae and other bones clearly in the scans that Burnham examines. Tardigrades on Earth are invertebrates with exoskeletons similar to those of insects. However, unlike insects or vertebrates, tardigrades do not have joints in their legs, but we can see in the episode that Ripper does. Ripper clearly has lungs (or equivalent structures) in that it breathes, roars, and has an audible "distress cry" in air. No Earth tardigrade could do this—they are water dwelling, and they respire by absorbing dissolved oxygen from their exterior surfaces. Ripper has the ability to function as a "supercomputer" in the episode. Tardigrade species on Earth possess a relatively simple brain structure, seemingly like the primitive ancestors of various modern-day species like insects.[22] Finally, and most obviously—as noted in the episode, tardigrades on Earth are less than a millimeter in length. Expanding to the size of Ripper would necessarily require massive changes in structure. Physics shows that one cannot simply make the same organism thousands of times larger by just making all component parts that much larger as well: it simply would not work. One might argue that physical limitation is why the other differences discussed above were present in Ripper relative to Earth tardigrades: to get around problems of large size. But the extent of these changes makes Ripper quite unlike a tardigrade in the end.

Why do I not label this as a "coal"? I still applaud the writers' creativity and their attempt to incorporate a recent scientific result into the episode. While later research disproved the tardigrade horizontal gene transfer result a little over a year before the episode aired, the writers may have already developed their script (or at least the basics of the storyline) before that time. I also appreciate their introduction of horizontal gene transfer into these and other episodes of the series, given the continuing scientific interest in its frequency and importance in biology.

NOTES

CHAPTER 1. "TO SEEK OUT NEW LIFE . . ."

1. Russell P, Hertz P, McMillan B. *Biology: The Dynamic Science*. Cengage Learning; 2011.

2. Schrödinger E. *What Is Life? The Physical Aspect of the Living Cell*. Cambridge University Press; 1944.

3. Erez Z, Steinberger-Levy I, Shamir M, et al. Communication between viruses guides lysis-lysogeny decisions. *Nature*. 2017;541(7638):488–493. doi:10.1038/nature21049.

4. Koonin EV, Starokadomskyy P. Are viruses alive? The replicator paradigm sheds decisive light on an old but misguided question. *Stud Hist Philos Biol Biomed Sci*. 2016;59:125–134. doi:10.1016/j.shpsc.2016.02.016. Leslie M. Cell-like giant viruses found. *Science*. 2017;356(6333):15–16. doi:10.1126/science.356.6333.15.

5. Szostak JW. Attempts to define life do not help to understand the origin of life. *J Biomol Struct Dyn*. 2012;29(4):599–600. doi:10.1080/073911012010524998.

6. Trifonov EN. Vocabulary of definitions of life suggests a definition. *J Biomol Struct Dyn*. 2011;29(2):259–266. doi:10.1080/073911011010524992.

7. Lincoln TA, Joyce GF. Self-sustained replication of an RNA enzyme. *Science*. 2009;323(5918):1229–1232. doi:10.1126/science.1167856. Petkovic S, Müller S. RNA self-processing: Formation of cyclic species and concatemers from a small engineered RNA. *FEBS Lett*. 2013;587(15):2435–2440. doi:10.1016/j.febslet.2013.06.013.

8. Horning DP, Joyce GF. Amplification of RNA by an RNA polymerase ribozyme. *Proc Natl Acad Sci USA*. 2016;113(35):9786–9791. doi:10.1073/pnas.1610103113.

9. Gross M. How life can arise from chemistry. *Curr Biol*. 2016;26(24):R1247–R1249. doi:10.1016/j.cub.2016.12.001. Pressman A, Blanco C, Chen IA. The RNA world as a model system to study the origin of life. *Curr Biol*. 2015;25(19):R953–R963. doi:10.1016/j.cub.2015.06.016.

10. Szostak JW. On the origin of life. *Medicina (B Aires)*. 2016;76(4):199–203. https://www.ncbi.nlm.nih.gov/pubmed/27576276.

11. Miller SL. A production of amino acids under possible primitive Earth conditions. *Science*. 1953;117(3046):528–529. doi:10.1126/science.117.3046.528.

12. Pearce BKD, Pudritz RE, Semenov DA, Henning TK. Origin of the RNA world: The fate of nucleobases in warm little ponds. *Proc Natl Acad Sci USA*. 2017;114(43):11327–11332. doi:10.1073/pnas.1710339114.

13. Bada JL. New insights into prebiotic chemistry from Stanley Miller's spark discharge experiments. *Chem Soc Rev*. 2013;42(5):2186–2196. doi:10.1039/C3CS35433D.

14. England JL. Statistical physics of self-replication. *J Chem Phys*. 2013; 139(12):121923. doi:10.1063/1.4818538.

15. Borucki WJ, Koch DG, Batalha N, et al. Kepler 22b: A 2.4 Earth-radius planet in the habitable zone of a sun-like star. *Astrophys J*. 2012;745(2):120. doi:10.1088/0004–637X/745/2/120. Borucki WJ, Agol E, Fressin F, et al. Kepler-62: A five-planet system with planets of 1.4 and 1.6 Earth radii in the habitable zone. *Science*. 2013;340(6132):587–590. doi:10.1126/science.1234702. Jenkins JM, Twicken JD. Batalha NM, et al. Discovery and validation of Kepler-452b: A 1.6 R super Earth exoplanet in the habitable zone of a star. *Astronom J*. 2015;150(2):56. doi:10.1088/0004-6256/150/2/56.

16. Gillon M, Triaud AH, Demory BO, et al. Seven temperate terrestrial planets around the nearby ultracool dwarf star TRAPPIST-1. *Nature*. 2017;542 (7642):456–460. doi:10.1038/nature21360.

17. Dittmann JA, Irwin JM, Charbonneau D, et al. A temperate rocky super-Earth transiting a nearby cool star. *Nature*. 2017;544(7650):333–336. doi:10.1038/nature22055.

18. Pohorille A, Pratt LR. Is water the universal solvent for life? *Orig Life Evol Biosph*. 2012;42(5):405–409. doi:10.1007/s11084-012-9301-6.

19. National Research Council. *The Limits of Organic Life in Planetary Systems*. National Academies Press; 2007.

20. Tobie G, Lunine JI, Sotin C. Episodic outgassing as the origin of atmospheric methane on Titan. *Nature*. 2006;440(7080):61–64. doi:10.1038/nature 04497.

21. Budisa N, Schulze-makuch D. Supercritical carbon dioxide and its potential as a life-sustaining solvent in a planetary environment. *Life (Basel)*. 2014;4(3):331–340. doi:10.3390/life4030331.

22. Stevenson J, Lunine J, Clancy P. Membrane alternatives in worlds without oxygen: Creation of an azotosome. *Sci Adv*. 2015;1(1):e1400067. doi:10.1126/sciadv.1400067.

23. Asimov I. Not as we know it: The chemistry of life. *Cosmic Search*. 1981; 3(1):5. http://www.bigear.org/CSMO/HTML/CS09/cs09p05.htm.

24. Ruiz-mirazo K, Briones C, De la Escosura A. Prebiotic systems chemistry: New perspectives for the origins of life. *Chem Rev*. 2014;114(1):285–366. doi:10.1021%2Fcr2004844.

25. Engel MH, Macko SA. Isotopic evidence for extraterrestrial non-racemic amino acids in the Murchison meteorite. *Nature*. 1997;389(6648):265–268. doi:10.1038/38460.

26. Cooper G, Kimmich N, Belisle W, Sarinana J, Brabham K, Garrel L. Carbonaceous meteorites as a source of sugar-related organic compounds for the early Earth. *Nature.* 2001;414(6866):879–883. doi:10.1038/414879a. Callahan MP, Smith KE, Cleaves HJ, et al. Carbonaceous meteorites contain a wide range of extraterrestrial nucleobases. *Proc Natl Acad Sci USA.* 2011;108(34):13995–13998. doi:10.1073/pnas.1106493108.

27. Kan SB, Lewis RD, Chen K, Arnold FH. Directed evolution of cytochrome c for carbon-silicon bond formation: Bringing silicon to life. *Science.* 2016;354(6315):1048–1051. doi:10.1126/science.aah6219.

28. Sleep NH, Zahnle K, Neuhoff PS. Initiation of clement surface conditions on the earliest Earth. *Proc Natl Acad Sci USA.* 2001;98(7):3666–3672. doi:10.1073/pnas.071045698.

29. Fiala G, Stetter KO. *Pyrococcus furiosus* sp. nov. represents a novel genus of marine heterotrophic archaebacteria growing optimally at 100°C. *Arch Microbiol.* 1986;145(1):56–61. doi:10.1007/BF00413027.

30. Kashefi K, Lovley DR. Extending the upper temperature limit for life. *Science.* 2003;301(5635):934. doi:10.1126/science.1086823.

31. Xu Y, Nogi Y, Kato C, et al. *Moritella profunda* sp. nov. and *Moritella abyssi* sp. nov., two psychropiezophilic organisms isolated from deep Atlantic sediments. *Int J Syst Evol Microbiol.* 2003;53(2):533–538. doi:10.1099/ijs.0.02228-0.

32. Clarke A, Morris GJ, Fonseca F, Murray BJ, Acton E, Price HC. A low temperature limit for life on Earth. *PLoS ONE.* 2013;8(6):e66207. doi:10.1371/journal.pone.0066207.

33. Persson D, Halberg KA, Jørgensen A, Ricci C, Møbjerg N, Kristensen RM. Extreme stress tolerance in tardigrades: Surviving space conditions in low Earth orbit. *J Zool Systemat Evol Res.* 2011;49(s1):90–97. doi:10.1111/j.1439-0469.2010.00605.x.

34. Jönsson KI, Rabbow E, Schill RO, Harms-ringdahl M, Rettberg P. Tardigrades survive exposure to space in low Earth orbit. *Curr Biol.* 2008;18(17):R729–R731. doi:10.1016/j.cub.2008.06.048.

35. Jönsson KI, Bertolani R. Facts and fiction about long-term survival in tardigrades. *J Zool.* 2001;255(1):121–123. doi:10.1017/S0952836901001169.

36. Hoover RB, Pikuta EV. Life in ice: Implications to astrobiology. *Proc SPIE.* 2009;7441. doi:10.1117/12.832640.

37. Proctor LM. Nitrogen-fixing, photosynthetic, anaerobic bacteria associated with pelagic copepods. *Aquat Microb Ecol.* 1997;12:105–113. doi:10.3354/ame012105.

38. Beatty JT, Overmann J, Lince MT, et al. An obligately photosynthetic bacterial anaerobe from a deep-sea hydrothermal vent. *Proc Natl Acad Sci USA.* 2005;102(26):9306–9310. doi:10.1073/pnas.0503674102.

39. Dadachova E, Bryan RA, Huang X, et al. Ionizing radiation changes the electronic properties of melanin and enhances the growth of melanized fungi. *PLoS ONE.* 2007;2(5):e457. doi:10.1371/journal.pone.0000457.

40. Boddy J. Alien life could feed on cosmic rays. *Science.* October 7, 2016. doi:10.1126/science.aal0237.

41. Atri D. On the possibility of galactic cosmic ray-induced radiolysis-powered life in subsurface environments in the Universe. *J R Soc Interface*. 2016; 13(123):20160459. doi:10.1098/rsif.2016.0459.

42. Gregory KB, Bond DR, Lovley DR. Graphite electrodes as electron donors for anaerobic respiration. *Environ Microbiol*. 2004;6(6):596–604. doi:10.1111/j.1462-2920.2004.00593.x.

43. Asimov I. Not as we know it: The chemistry of life. *Cosmic Search*. 1981; 3(1):5. http://www.bigear.org/CSMO/HTML/CS09/cs09p05.htm.

44. Morris CE, Sands DC, Bardin M, et al. Microbiology and atmospheric processes: Research challenges concerning the impact of airborne micro-organisms on the atmosphere and climate. *Biogeosciences*. 2011;8:17–25. doi:10.5194/bg-8-17-2011.

CHAPTER 2. CHARTING THE RELATIONSHIPS OF SPECIES

1. Darwin C. *On the Origin of Species By Means of Natural Selection or the Preservation of Favoured Races in the Struggle For Life*. D. Appleton; 1861:425.

2. Theobald DL. A formal test of the theory of universal common ancestry. *Nature*. 2010;465(7295):219–222. doi:10.1038/nature09014.

3. Yonezawa T, Hasegawa M. Was the universal common ancestry proved? *Nature*. 2010;468(7326):E9. doi:10.1038/nature09482.

4. Linnaei C. *Systema Naturae*. Lugduni Batavorum; 1735. http://www.biodiversitylibrary.org/item/15373.

5. Cohn MJ, Tickle C. Developmental basis of limblessness and axial patterning in snakes. *Nature*. 1999;399(6735):474–479. doi:10.1038/20944.

6. Li Y, Shi CX, Mossman KL, Rosenfeld J, Boo YC, Schellhorn HE. Restoration of vitamin C synthesis in transgenic Gulo–/– mice by helper-dependent adenovirus-based expression of gulonolactone oxidase. *Hum Gene Ther*. 2008; 19(12):1349–1358. doi:10.1089/hgt.2008.106.

7. Caldwell MW, Lee MSY. A snake with legs from the marine Cretaceous of the Middle East. *Nature*. 1997;386:705–709. doi:10.1038/386705a0. Apesteguía S, Zaher H. A Cretaceous terrestrial snake with robust hindlimbs and a sacrum. *Nature*. 2006;440(7087):1037–1040.doi:10.1038/nature04413.

8. Xu X, Zhou Z, Dudley R, et al. An integrative approach to understanding bird origins. *Science*. 2014;346(6215):1253293. doi:10.1126/science.1253293.

9. Thewissen JG, Cooper LN, Clementz MT, Bajpai S, Tiwari BN. Whales originated from aquatic artiodactyls in the Eocene epoch of India. *Nature*. 2007; 450(7173):1190–1194. doi:10.1038/nature06343.

10. Bettridge S, Baker CS, Barlow J, et al. *Status Review of the Humpback Whale (Megaptera novaeangliae) under the Endangered Species Act*. NOAA-TM-NMFS-SWFSC-540. US Department of Commerce; 2015. https://swfsc.noaa.gov/publications/TM/SWFSC/NOAA-TM-NMFS-SWFSC-540.pdf.

11. Engel MH, Macko SA. Isotopic evidence for extraterrestrial non-racemic amino acids in the Murchison meteorite. *Nature*. 1997;389(6648):265–268. doi:10.1038/38460. Cooper G, Kimmich N, Belisle W, Sarinana J, Brabham K, Garrel L. Carbonaceous meteorites as a source of sugar-related organic compounds for the early Earth. *Nature*. 2001;414(6866):879–883. doi:10.1038/414879a.

12. Elsila JE, Glavin DP, Dworkin JP. Cometary glycine detected in samples returned by stardust. *Meteor Planet Sci*. 2009;44(9)1323–1330. doi:10.1111/j.1945–5100.2009.tb01224.x.

13. Nicholson WL. Ancient micronauts: Interplanetary transport of microbes by cosmic impacts. *Trends Microbiol*. 2009;17(6):243–250. doi:10.1016/j.tim.2009.03.004. Horneck G, Klaus DM, Mancinelli RL. Space microbiology. *Microbiol Mol Biol Rev*. 2010;74(1):121–156. doi:10.1128/MMBR.00016–09.

14. Lee H, Popodi E, Tang H, Foster PL. Rate and molecular spectrum of spontaneous mutations in the bacterium *Escherichia coli* as determined by whole-genome sequencing. *Proc Natl Acad Sci USA*. 2012;109(41):E2774–E2783. doi:10.1073/pnas.1210309109.

15. Howard BC. What are the odds a meteorite could kill you? *National Geographic*. February 9, 2016. http://news.nationalgeographic.com/2016/02/160209-meteorite-death-india-probability-odds/.

16. Gould SJ. *Wonderful Life: The Burgess Shale and the Nature of History*. W. W. Norton; 1990.

17. Ibid., 318.

18. Poole AM, Gribaldo S. Eukaryotic origins: How and when was the mitochondrion acquired? *Cold Spring Harb Perspect Biol*. 2014;6(12):a015990. doi:10.1101/cshperspect.a015990.

19. Margulis L. *Symbiosis in Cell Evolution: Life and its Environment on the Early Earth*. W. H. Freeman; 1981. O'Malley MA. Endosymbiosis and its implications for evolutionary theory. *Proc Natl Acad Sci USA*. 2015;112(33):10270–10277. doi:10.1073/pnas.1421389112.

20. For an earlier example, see: Serviss GP, Godwin R. *Edison's Conquest of Mars*. Apogee; 2010.

21. Sagan C. *Broca's Brain: Reflections on the Romance of Science*. Presidio; 1980.

22. Goren-inbar N, Sharon G, Melamed Y, Kislev M. Nuts, nut cracking, and pitted stones at Gesher Benot Ya'aqov, Israel. *Proc Natl Acad Sci USA*. 2002;99(4):2455–2460. doi:10.1073/pnas.032570499. Ben-dor M, Gopher A, Hershkovitz I, Barkai R. Man the fat hunter: The demise of *Homo erectus* and the emergence of a new hominin lineage in the Middle Pleistocene (ca. 400 kyr) Levant. *PLoS ONE*. 2011;6(12):e28689. doi:10.1371/journal.pone.0028689.

23. Goren-inbar N, Alperson N, Kislev ME, et al. Evidence of hominin control of fire at Gesher Benot Ya'aqov, Israel. *Science*. 2004;304(5671):725–727. doi:10.1126/science.1095443. Roebroeks W, Villa P. On the earliest evidence for habitual use of fire in Europe. *Proc Natl Acad Sci USA*. 2011;108(13):5209–5214. doi:10.1073/pnas.1018116108.

24. Harrison T. Anthropology: Apes among the tangled branches of human origins. *Science*. 2010;327(5965):532–534. doi:10.1126/science.1184703.

25. Young NM, Capellini TD, Roach NT, Alemseged Z. Fossil hominin shoulders support an African ape-like last common ancestor of humans and chimpanzees. *Proc Natl Acad Sci USA.* 2015;112(38):11829–11834. doi:10.1073/pnas.1511220112. Almécija S. Pitfalls reconstructing the last common ancestor of chimpanzees and humans. *Proc Natl Acad Sci USA.* 2016;113(8):E943–E944. doi:10.1073/pnas.1524165113.

26. Chimpanzee Sequencing and Analysis Consortium. Initial sequence of the chimpanzee genome and comparison with the human genome. *Nature.* 2005;437(7055):69–87. doi:10.1038/nature04072.

27. Fairén AG, Schulze-Makuch D. The overprotection of Mars. *Nature Geosci.* 2013;6:510–511. doi:10.1038/ngeo1866.

28. Mileikowsky C, Cucinotta FA, Wilson JW, et al. Natural transfer of viable microbes in space: 1. From Mars to Earth and Earth to Mars. *Icarus.* 2000;145(2)391–427. doi:10.1006/icar.1999.6317.

29. NASA. *Planetary Protection for Mars Science Laboratory.* NASA Facts. NASA; 2012. https://www.jpl.nasa.gov/news/fact_sheets/MSLPlanProtFact.pdf.

30. Frick A, Mogul R, Stabekis P, Conley CA, Ehrenfreund P. Overview of current capabilities and research and technology developments for planetary protection. *Adv Space Res.* 2014;54(2)221–240. doi:10.1016/j.asr.2014.02.016.

31. Scaduto DI, Brown JM, Haaland WC, Zwickl DJ, Hillis DM, Metzker ML. Source identification in two criminal cases using phylogenetic analysis of HIV-1 DNA sequences. *Proc Natl Acad Sci USA.* 2010;107(50):21242–21247. doi:10.1073/pnas.1015673107.

32. Fassett JE. New geochronologic and stratigraphic evidence confirms the Paleocene age of the dinosaur-bearing Ojo Alamo Sandstone and Animas Formation in the San Juan Basin, New Mexico and Colorado. *Palaeontol Electron.* 2009:12(3A):1–146. http://palaeo-electronica.org/2009_1/149/index.html.

33. Lucas SG, Sullivan RM, Cather SM, et al. No definitive evidence of Paleocene dinosaurs in the San Juan Basin. *Palaeontol Electron.* 2009;12(8A):1–10. Renne PR, Goodwin MB. Direct U-Pb dating of Cretaceous and Paleocene dinosaur bones, San Juan Basin, New Mexico: COMMENT. *Geology.* 2012;40(4)e259. doi:10.1130/G32521C.1. Brusatte SL, Butler RJ, Barrett PM, et al. The extinction of the dinosaurs. *Biol Rev Camb Philos Soc.* 2015;90(2):628–642. doi:10.1111/brv.12128.

CHAPTER 3. DNA: EVOLUTION'S CAPTAIN

1. O'Leary NA, Wright MW, Brister JR, et al. Reference sequence (RefSeq) database at NCBI: Current status, taxonomic expansion, and functional annotation. *Nucleic Acids Res.* 2016;44(D1):D733–D745. doi:10.1093/nar/gkv1189.

2. Altschul SF, Gish W, Miller W, Myers EW, Lipman DJ. Basic local alignment search tool. *J Mol Biol.* 1990;215(3):403–410. doi:10.1016/S0022-2836(05)80360-2.

3. Wu R. Nucleotide sequence analysis of DNA. *Nat New Biol*. 1972;236:198–200. doi:10.1038/newbio236198a0.

4. International Human Genome Sequencing Consortium. Initial sequencing and analysis of the human genome. *Nature*. 2001;409(6822):860–921. doi:10.1038/35057062. Venter JC, Adams MD, Myers EW, et al. The sequence of the human genome. *Science*. 2001;291(5507):1304–1351. doi:10.1126/science.1058040.

5. Avery OT, MacLeod CM, McCarty M. Studies of the chemical nature of the substance inducing transformation of pneumococcal types. *J Exp Med*. 1944;79(2):137–159. doi:10.1084/jem.79.2.137. Hershey AD, Chase M. Independent functions of viral protein and nucleic acid in growth of bacteriophage. *J Gen Physiol*. 1952;36(1):39–56. doi:10.1085/jgp.36.1.39.

6. Crick F. Central dogma of molecular biology. *Nature*. 1970;227:561–563. doi:10.1038/227561a0. Cobb M. 60 years ago, Francis Crick changed the logic of biology. *PLoS Biol*. 2017;15(9):e2003243. doi:10.1371/journal.pbio.2003243.

7. Zhang R, Lahens NF, Ballance HI, Hughes ME, Hogenesch JB. A circadian gene expression atlas in mammals: Implications for biology and medicine. *Proc Natl Acad Sci USA*. 2014;111(45):16219–16224. doi:10.1073/pnas.1408886111.

8. Pan Q, Shai O, Lee LJ, Frey BJ, Blencowe BJ. Deep surveying of alternative splicing complexity in the human transcriptome by high-throughput sequencing. *Nat Genet*. 2008;40(12):1413–1415. doi:10.1038/ng.259. Wang ET, Sandberg R, Luo S, et al. Alternative isoform regulation in human tissue transcriptomes. *Nature*. 2008;456(7221):470–476. doi:10.1038/nature07509.

9. Lee CK, Klopp RG, Weindruch R, Prolla TA. Gene expression profile of aging and its retardation by caloric restriction. *Science*. 199;285(5432):1390–1393. doi:10.1126/science.285.5432.1390.

10. Mitchell HK, Petersen NS. The recessive phenotype of forked can be uncovered by heat shock in *Drosophila*. *Genesis*. 1985;6(2):93–100. doi:10.1002/dvg.1020060203.

11. Tomasetti C, Vogelstein B. Cancer etiology: Variation in cancer risk among tissues can be explained by the number of stem cell divisions. *Science*. 2015;347(6217):78–81. doi:10.1126/science.1260825.

12. Owens B. Genomics: The single life. *Nature*. 2012;491(7422):27–29. doi:10.1038/491027a.

13. Lawrence MS, Stojanov P, Polak P, et al. Mutational heterogeneity in cancer and the search for new cancer-associated genes. *Nature*. 2013;499(7457):214–218. doi:10.1038/nature12213.

14. Roach JC, Glusman G, Smit AF, et al. Analysis of genetic inheritance in a family quartet by whole-genome sequencing. *Science*. 2010;328(5978):636–639. doi:10.1126/science.1186802.

15. Wang J, Fan HC, Behr B, Quake SR. Genome-wide single-cell analysis of recombination activity and de novo mutation rates in human sperm. *Cell*. 2012;150(2):402–412. doi:10.1016/j.cell.2012.06.030.

16. Deans C, Maggert KA. What do you mean, "epigenetic"? *Genetics*. 2015;199(4):887–896. doi:10.1534/genetics.114.173492.

17. Barrès R, Yan J, Egan B, et al. Acute exercise remodels promoter methylation in human skeletal muscle. *Cell Metab.* 2012;15(3):405–411. doi:10.1016/j .cmet.2012.01.001.

18. Jones MJ, Goodman SJ, Kobor MS. DNA methylation and healthy human aging. *Aging Cell.* 2015;14(6):924–932. doi:10.1111/acel.12349.

19. Baker EK, Johnstone RW, Zalcberg JR, El-osta A. Epigenetic changes to the MDR1 locus in response to chemotherapeutic drugs. *Oncogene.* 2005;24(54): 8061–8075. doi:10.1038/sj.onc.1208955.

20. Morgan HD, Sutherland HG, Martin DI, Whitelaw E. Epigenetic inheritance at the agouti locus in the mouse. *Nat Genet.* 1999;23(3):314–318. doi:10.1038/15490. Rakyan VK, Chong S, Champ ME, et al. Transgenerational inheritance of epigenetic states at the murine Axin(Fu) allele occurs after maternal and paternal transmission. *Proc Natl Acad Sci USA.* 2003;100(5):2538–2543. doi:10.1073/pnas.0436776100.

21. Heard E, Martienssen RA. Transgenerational epigenetic inheritance: Myths and mechanisms. *Cell.* 2014;157(1):95–109. doi:10.1016/j.cell.2014.02 .045. Daxinger L, Whitelaw E. Understanding transgenerational epigenetic inheritance via the gametes in mammals. *Nat Rev Genet.* 2012;13(3):153–162. doi:10 .1038/nrg3188.

22. Ng SF, Lin RC, Laybutt DR, Barres R, Owens JA, Morris MJ. Chronic high-fat diet in fathers programs β-cell dysfunction in female rat offspring. *Nature.* 2010;467(7318):963–966. doi:10.1038/nature09491.

23. Van Otterdijk SD, Michels KB. Transgenerational epigenetic inheritance in mammals: How good is the evidence? *FASEB J.* 2016;30(7):2457–2465. doi:10.1096/fj.201500083. Charlesworth D, Barton NH, Charlesworth B. The sources of adaptive variation. *Proc Biol Sci.* 2017;284(1855):20162864. doi:10.1098 /rspb.2016.2864.

24. Bergmann O, Zdunek S, Felker A, et al. Dynamics of cell generation and turnover in the human heart. *Cell.* 2015;161(7):1566–1575. doi:10.1016/j .cell.2015.05.026.

25. Angers B, Castonguay E, Massicotte R. Environmentally induced phenotypes and DNA methylation: How to deal with unpredictable conditions until the next generation and after. *Mol Ecol.* 2010;19(7):1283–1295. doi:10.1111/j.1365 –294X.2010.04580.x.

26. Moreno-Villanueva M, Wong M, Lu T, Zhang Y, Wu H. Interplay of space radiation and microgravity in DNA damage and DNA damage response. *NPJ Microgravity.* 2017;3:14. doi:10.1038/s41526-017-0019-7.

27. Andersen SL, Sekelsky J. Meiotic versus mitotic recombination: Two different routes for double-strand break repair; the different functions of meiotic versus mitotic DSB repair are reflected in different pathway usage and different outcomes. *Bioessays.* 2010;32(12):1058–1066. doi:10.1002/bies.201000087.

28. Kavanagh JN, Currell FJ, Timson DJ, et al. Antiproton induced DNA damage: Proton like in flight, carbon-ion like near rest. *Sci Rep.* 2013;3:1770. doi:10.1038/srep01770.

29. Tsukuda T, Fleming AB, Nickoloff JA, Osley MA. Chromatin remodelling at a DNA double-strand break site in *Saccharomyces cerevisiae*. *Nature*. 2005;438 (7066):379–383. doi:10.1038/nature04148.

30. Hunt CR, Ramnarain D, Horikoshi N, et al. Histone modifications and DNA double-strand break repair after exposure to ionizing radiations. *Radiat Res*. 2013;179(4):383–392. doi:10.1667/RR3308.2.

31. Jinek M, Chylinski K, Fonfara I, Hauer M, Doudna JA, Charpentier E. A programmable dual-RNA-guided DNA endonuclease in adaptive bacterial immunity. *Science*. 2012;337(6096):816–821. doi:10.1126/science.1225829.

32. Tang L, Zeng Y, Du H, et al. CRISPR/Cas9-mediated gene editing in human zygotes using Cas9 protein. *Mol Genet Genomics*. 2017;292(3):525–533. doi:10.1007/s00438-017-1299-z.

CHAPTER 4. CHANGE OVER TIME: DRIVERS OF EVOLUTION

1. Mead R, Hejmadi M, Hurst LD. Teaching genetics prior to teaching evolution improves evolution understanding but not acceptance. *PLoS Biol*. 2017; 15(5):e2002255. doi:10.1371/journal.pbio.2002255.

2. Eory L, Halligan DL, Keightley PD. Distributions of selectively constrained sites and deleterious mutation rates in the hominid and murid genomes. *Mol Biol Evol*. 2010;27(1):177–192. doi:10.1093/molbev/msp219. Keightley PD. Rates and fitness consequences of new mutations in humans. *Genetics*. 2012;190(2): 295–304. doi:10.1534/genetics.111.134668.

3. Darwin, C. *On the Origin of Species by Means of Natural Selection, or Preservation of Favoured Races in the Struggle for Life*. John Murray; 1859.

4. Burger J, Kirchner M, Bramanti B, Haak W, Thomas MG. Absence of the lactase-persistence-associated allele in early Neolithic Europeans. *Proc Natl Acad Sci USA*. 2007;104(10):3736–3741. doi:10.1073/pnas.0607187104.

5. Itan Y, Powell A, Beaumont MA, Burger J, Thomas MG. The origins of lactase persistence in Europe. *PLoS Comput Biol*. 2009;5(8):e1000491. doi:10.1371 /journal.pcbi.1000491.

6. Tishkoff SA, Reed FA, Ranciaro A, et al. Convergent adaptation of human lactase persistence in Africa and Europe. *Nat Genet*. 2007;39(1):31–40. doi:10.1038 /ng1946.

7. Purugganan MD, Boyles AL, Suddith JI. Variation and selection at the cauliflower floral homeotic gene accompanying the evolution of domesticated *Brassica oleracea*. *Genetics*. 2000;155(2):855–862.

8. Chambers HF, Deleo FR. Waves of resistance: *Staphylococcus aureus* in the antibiotic era. *Nat Rev Microbiol*. 2009;7(9):629–641. doi:10.1038/nrmicro2200.

9. Drury B, Scott J, Rosi-marshall EJ, Kelly JJ. Triclosan exposure increases triclosan resistance and influences taxonomic composition of benthic bacterial communities. *Environ Sci Technol*. 2013;47(15):8923–8930. doi:10.1021/es401919k.

10. Durand R, Bouvresse S, Berdjane Z, Izri A, Chosidow O, Clark JM. Insecticide resistance in head lice: Clinical, parasitological and genetic aspects. *Clin Microbiol Infect.* 2012;18(4):338–344. doi:10.1111/j.1469–0691.2012.03806.x.

11. Leng J, Goldstein DR. Impact of aging on viral infections. *Microb Infect.* 2010;12(14–15):1120–1124. doi:10.1016/j.micinf.2010.08.009.

12. Monaghan P. Telomeres and life histories: The long and the short of it. *Ann NY Acad Sci.* 2010;1206:130–142. doi:10.1111/j.1749–6632.2010.05705.x.

13. Dawkins R. *The Blind Watchmaker: Why the Evidence of Evolution Reveals a Universe without Design.* W. W. Norton; 1986.

14. Dumont ER. Bone density and the lightweight skeletons of birds. *Proc Biol Sci.* 2010;277(1691):2193–2198. doi:10.1098/rspb.2010.0117.

15. Patek SN, Korff WL, Caldwell RL. Biomechanics: Deadly strike mechanism of a mantis shrimp. *Nature.* 2004;428(6985):819–820. doi:10.1038/428819a. Patek SN. Extreme power output and ultrafast movements in biology. *Am Sci,* 2015;103(5):330–337.

16. Liu H, Ravi S, Kolomenskiy D, Tanaka H. Biomechanics and biomimetics in insect-inspired flight systems. *Philos Trans R Soc Lond, B.* 2016;371(1704): 20150390. doi:10.1098/rstb.2015.0390. Goldin DS, Venneri SL, Noor AK. The great out of the small. *Mech Eng.* 2000;122(11):70–79.

17. Skoglund P, Ersmark E, Palkopoulou E, Dalén L. Ancient wolf genome reveals an early divergence of domestic dog ancestors and admixture into high-latitude breeds. *Curr Biol.* 2015;25(11):1515–1519. doi:10.1016/j.cub.2015.04.019.

18. Rubinstein CV, Gerrienne P, De la Puente GS, Astini RA, Steemans P. Early Middle Ordovician evidence for land plants in Argentina (eastern Gondwana). *New Phytol.* 2010;188(2):365–369. doi:10.1111/j.1469–8137.2010.03433.x.

19. Gingerich PD, Wells NA, Russell DE, Shah SM. Origin of whales in epicontinental remnant seas: New evidence from the early Eocene of Pakistan. *Science.* 1983;220(4595):403–406. doi:10.1126/science.220.4595.403. Thewissen JG, Williams EM, Roe LJ, Hussain ST. Skeletons of terrestrial cetaceans and the relationship of whales to artiodactyls. *Nature.* 2001;413(6853):277–281. doi:10.1038/35095005.

20. Gould SJ. *The Panda's Thumb: More Reflections in Natural History.* W. W. Norton; 1992.

21. Mahler DL, Weber MG, Wagner CE, Ingram T. Pattern and process in the comparative study of convergent evolution. *Am Nat.* 2017;190(S1):S13–S28. doi:10.1086/692648.

22. Charlesworth D, Willis JH. The genetics of inbreeding depression. *Nat Rev Genet.* 2009;10(11):783–796. doi:10.1038/nrg2664.

23. Branda KJ, Tomczak J, Natowicz MR. Heterozygosity for Tay–Sachs and Sandhoff diseases in non-Jewish Americans with ancestry from Ireland, Great Britain, or Italy. *Genet Test.* 2004;8(2):174–180. doi:10.1089/gte.2004.8.174.

24. Hayden MR, Berkowicz AL, Beighton PH, Yiptong C. Huntington's chorea on the island of Mauritius. *S Afr Med J.* 1981;60(26):1001–1002.

25. Pringsheim T, Wiltshire K, Day L, Dykeman J, Steeves T, Jette N. The incidence and prevalence of Huntington's disease: A systematic review and meta-analysis. *Mov Disord.* 2012;27(9):1083–1091. doi:10.1002/mds.25075.

26. Kong A, Frigge ML, Masson G, et al. Rate of de novo mutations and the importance of father's age to disease risk. *Nature.* 2012;488(7412):471–475. doi:10.1038/nature11396.

27. Lindblad-toh K, Garber M, Zuk O, et al. A high-resolution map of human evolutionary constraint using 29 mammals. *Nature.* 2011;478(7370):476–482. doi:10.1038/nature10530.

28. Bigham A, Bauchet M, Pinto D, et al. Identifying signatures of natural selection in Tibetan and Andean populations using dense genome scan data. *PLoS Genet.* 2010;6(9):e1001116. doi:10.1371/journal.pgen.1001116. Yi X, Liang Y, Huerta-Sanchez E, et al. Sequencing of 50 human exomes reveals adaptation to high altitude. *Science.* 2010;329(5987):75–78. doi:10.1126/science.1190371. Huerta-Sánchez E, Degiorgio M, Pagani L, et al. Genetic signatures reveal high-altitude adaptation in a set of Ethiopian populations. *Mol Biol Evol.* 2013;30(8):1877–1888. doi:10.1093/molbev/mst089.

29. Li H, Durbin R. Inference of human population history from individual whole-genome sequences. *Nature.* 2011;475(7357):493–496. doi:10.1038/nature10231. Keinan A, Clark AG. Recent explosive human population growth has resulted in an excess of rare genetic variants. *Science.* 2012;336(6082):740–743. doi:10.1126/science.1217283.

30. Bittles AH, Black ML. Consanguinity, human evolution, and complex diseases. *Proc Natl Acad Sci USA.* 2010;107(Suppl 1):1779–1786. doi:10.1073/pnas.0906079106. Pemberton TJ, Absher D, Feldman MW, Myers RM, Rosenberg NA, Li JZ. Genomic patterns of homozygosity in worldwide human populations. *Am J Hum Genet.* 2012;91(2):275–292. doi:10.1016/j.ajhg.2012.06.014.

CHAPTER 5. SEX, REPRODUCTION, AND THE MAKING OF NEW SPECIES

1. Redfield RJ. Do bacteria have sex? *Nat Rev Genet.* 2001;2(8):634–639. doi:10.1038/35084593.

2. Barton NH, Charlesworth B. Why sex and recombination? *Science.* 1998; 281(5385):1986–1990. doi:10.1126/science.281.5385.1986. Butlin R. Evolution of sex: The costs and benefits of sex: new insights from old asexual lineages. *Nat Rev Genet.* 2002;3(4):311–317. doi:10.1038/nrg749. Hartfield M, Keightley PD. Current hypotheses for the evolution of sex and recombination. *Integr Zool.* 2012;7(2):192–209. doi:10.1111/j.1749-4877.2012.00284.x.

3. Otto SP. The evolutionary enigma of sex. *Am Nat.* 2009;174(Suppl 1):S1–S14. doi:10.1086/599084.

4. Schwander T, Crespi BJ. Twigs on the tree of life? Neutral and selective models for integrating macroevolutionary patterns with microevolutionary processes in the analysis of asexuality. *Mol Ecol.* 2009;18(1):28–42. doi:10.1111/j.1365-294X.2008.03992.x. Hartfield M, Keightley PD. Current hypotheses for the evolution of sex and recombination. *Integr Zool.* 2012;7(2):192–209. doi:10.1111/j.1749-4877.2012.00284.x.

5. Lodé T. Adaptive significance and long-term survival of asexual lineages. *Evol. Biol.* 2013;40(3)450–460. doi:10.1007/s11692-012-9219-y.

6. Mark Welch DB, Meselson M. Evidence for the evolution of bdelloid rotifers without sexual reproduction or genetic exchange. *Science.* 2000;288(5469):1211–1215. doi:10.1126/science.288.5469.1211. Mark Welch JL, Mark Welch DB, Meselson M. Cytogenetic evidence for asexual evolution of bdelloid rotifers. *Proc Natl Acad Sci USA.* 2004;101(6):1618–1621. doi:10.1073/pnas.0307677100. Gladyshev EA, Meselson M, Arkhipova IR. Massive horizontal gene transfer in bdelloid rotifers. *Science.* 2008;320(5880):1210–1213. doi:10.1126/science.1156407; Flot JF, Hespeels B, Li X, et al. Genomic evidence for ameiotic evolution in the bdelloid rotifer *Adineta vaga. Nature.* 2013;500(7463):453–457. doi:10.1038/nature12326.

7. Heitman J. Evolution of eukaryotic microbial pathogens via covert sexual reproduction. *Cell Host Microb.* 2010;8(1):86–99. doi:10.1016/j.chom.2010.06.011.

8. John Maynard Smith. Interview by *The Evolutionist*, conducted February 2, 1999. www.lse.ac.uk/CPNSS/research/projectsCurrentlyOnHold/darwin/publications/evolutionist/jms.aspx. Updated January 12, 2011. Accessed January 8, 2018.

9. Schlupp I, Riesch R, Tobler M. Amazon mollies. *Curr Biol.* 2007;17(14):R536–R537. doi:10.1016/j.cub.2007.05.012.

10. Pigneur LM, Hedtke SM, Etoundi E, Van Doninck K. Androgenesis: A review through the study of the selfish shellfish *Corbicula* spp. *Heredity (Edinb).* 2012;108(6):581–591. doi:10.1038/hdy.2012.3.

11. Schwander T, Oldroyd BP. Androgenesis: Where males hijack eggs to clone themselves. *Philos Trans R Soc Lond, B.* 2016;371(1706):20150534. doi:10.1098/rstb.2015.0534. Morgado-Santos M, Carona S, Vicente L, Collares-Pereira MJ. First empirical evidence of naturally occurring androgenesis in vertebrates. *R Soc Open Sci.* 2017;4(5):170200. doi:10.1098/rsos.170200.

12. Buston P. Does the presence of non-breeders enhance the fitness of breeders? An experimental analysis in the clown anemonefish *Amphiprion percula. Behav Ecol Sociobiol.* 2004;57(1):23–31. doi:10.1007/s00265-004-0833-2.

13. Hamilton WD. The genetical evolution of social behavior. I. *J Theor Biol.* 1964;7(1):1–16. doi:10.1016/0022-5193(64)90038-4.

14. Tuttle EM, Bergland AO, Korody ML, et al. Divergence and functional degradation of a sex chromosome-like supergene. *Curr Biol.* 2016;26(3):344–350. doi:10.1016/j.cub.2015.11.069. Arnold C. The sparrow with four sexes. *Nature.* 2016;539(7630):482–484. doi:10.1038/539482a.

15. Lloyd JE. Aggressive mimicry in *Photuris* fireflies: Signal repertoires by femmes fatales. *Science*. 1975;187(4175):452–453. doi:10.1126/science.187.4175 .452.

16. Bateman AJ. Intra-sexual selection in *Drosophila*. *Heredity*. 1948;2:349–368. doi:10.1038/hdy.1948.21. Knight J. Sexual stereotypes. *Nature*. 2002;415(6869): 254–256. doi:10.1038/415254a.

17. Honěk A. Intraspecific variation in body size and fecundity in insects: A general relationship. *Oikos*. 1993;66(3):483–492. doi:10.2307/3544943.

18. Dobzhansky T. A critique of the species concept in biology. *Philos Sci*. 1935;2(3):344–355. doi:10.1086/286379. Mayr E. *Systematics and the Origin of Species*. Columbia University Press; 1942.

19. Mallet J. Hybridization as an invasion of the genome. *Trends Ecol Evol (Amst)*. 2005;20(5):229–237. doi:10.1016/j.tree.2005.02.010.

20. Feder JL, Chilcote CA, Bush GL. Genetic differentiation between sympatric host races of the apple maggot fly *Rhagoletis pomonella*. *Nature*. 1988;336: 61–64. doi:10.1038/336061a0. Ragland GJ, Doellman MM, Meyers PJ, et al. A test of genomic modularity among life-history adaptations promoting speciation with gene flow. *Mol Ecol*. 2017;26(15):3926–3942. doi:10.1111/mec.14178.

21. Feder JL, Opp SB, Wlazlo B, Reynolds K, Go W, Spisak S. Host fidelity is an effective premating barrier between sympatric races of the apple maggot fly. *Proc Natl Acad Sci USA*. 1994;91(17):7990–7994.

22. Williams KS, Simon C. The ecology, behavior, and evolution of periodical cicadas. *Annu Rev Entomol*. 1995;40:269–295. doi:10.1146/annurev.en.40.010195 .001413.

23. Simon C, Tang J, Dalwadi S, Staley G, Deniega J, Unnasch TR. Genetic evidence for assortative mating between 13-year cicadas and sympatric "17-year cicadas with 13-year life cycles" provides support for allochronic speciation. *Evolution*. 2000;54(4):1326–1336. doi:10.1111/j.0014-3820.2000.tb00565.x.

24. Knowlton N, Maté JL, Guzmán HM, Rowan R, Jara J. Direct evidence for reproductive isolation among the three species of the *Montastraea annularis* complex in Central America (Panamá and Honduras). *Mar Biol*. 1997;127(4): 705–711. doi:10.1007/s002270050061. Levitan DR, Fukami H, Jara J, et al. Mechanisms of reproductive isolation among sympatric broadcast-spawning corals of the *Montastraea annularis* species complex. *Evolution*. 2004;58(2):308–323. doi:10.1554/02–700.

25. Stratron GE, Uetz GW. Acoustic communication and reproductive isolation in two species of wolf spiders. *Science*. 1981;214(4520):575–577. doi:10.1126 /science.214.4520.575.

26. Uetz GW, Roberts JA. Multisensory cues and multimodal communication in spiders: Insights from video/audio playback studies. *Brain Behav Evol*. 2002;59(4):222–230. doi:10.1159/000064909.

27. McClintock WJ, Uetz GW. Female choice and pre-existing bias: Visual cues during courtship in two *Schizocosa* wolf spiders (Araneae: Lycosidae). *Anim Behav*. 1996;52(1):167–181. doi:10.1006/anbe.1996.0162.

28. Coyne JA, Crittenden AP, Mah K. Genetics of a pheromonal difference contributing to reproductive isolation in *Drosophila*. *Science*. 1994;265(5177): 1461–1464. doi:10.1126/science.8073292.

29. Kresge N, Vacquier VD, Stout CD. Abalone lysin: The dissolving and evolving sperm protein. *Bioessays*. 2001;23(1):95–103. doi:10.1002/1521–1878(200101) 23:1<95::AID-BIES1012>3.0.CO;2-C.

30. Swanson WJ, Vacquier VD. The abalone egg vitelline envelope receptor for sperm lysin is a giant multivalent molecule. *Proc Natl Acad Sci USA*. 1997;94 (13):6724–6729.

31. Patterson JT. A new type of isolating mechanism in *Drosophila*. *Proc Natl Acad Sci USA*. 1946;32(7):202–208.

32. Bordenstein SR, O'Hara FP, Werren JH. *Wolbachia*-induced incompatibility precedes other hybrid incompatibilities in *Nasonia*. *Nature*. 2001;409 (6821):707–710. doi:10.1038/35055543.

33. Schartl M. Evolution of *Xmrk*: An oncogene, but also a speciation gene? *Bioessays*. 2008;30(9):822–832. doi:10.1002/bies.20807.

34. Rothfels CJ, Johnson AK, Hovenkamp PH, et al. Natural hybridization between genera that diverged from each other approximately 60 million years ago. *Am Nat*. 2015;185(3):433–442. doi:10.1086/679662.

35. Wodsedalek JE. Causes of sterility in the mule. *Biol Bull*. 1916;30(1):1–57. doi:10.2307/1536434.

36. Rong R, Chandley AC, Song J, et al. A fertile mule and hinny in China. *Cytogenet Cell Genet*. 1988;47(3):134–139. doi:10.1159/000132531.

37. Schilthuizen M, Giesbers MC, Beukeboom LW. Haldane's rule in the 21st century. *Heredity (Edinb)*. 2011;107(2):95–102. doi:10.1038/hdy.2010.170.

38. Haldane JBS. Sex ratio and unisexual sterility in hybrid animals. *J Genet*. 1922;12(2):101–109. doi:10.1007/BF02983075.

39. Brothers AN, Delph LF. Haldane's rule is extended to plants with sex chromosomes. *Evolution*. 2010;64(12):3643–3648. doi:10.1111/j.1558–5646.2010 .01095.x.

40. Delph LF, Demuth JP. Haldane's rule: Genetic bases and their empirical support. *J Hered*. 2016;107(5):383–391. doi:10.1093/jhered/esw026.

41. Sinclair KD, Corr SA, Gutierrez CG, et al. Healthy ageing of cloned sheep. *Nat Commun*. 2016;7:12359. doi:10.1038/ncomms12359.

42. Griffith SC, Owens IP, Thuman KA. Extra pair paternity in birds: A review of interspecific variation and adaptive function. *Mol Ecol*. 2002;11(11):2195–2212. doi:10.1046/j.1365–294X.2002.01613.x.

43. Mallet J. Hybridization as an invasion of the genome. *Trends Ecol Evol (Amst)*. 2005;20(5):229–237. doi:10.1016/j.tree.2005.02.010.

44. Abecasis GR, Altshuler D, Auton A, et al. A map of human genome variation from population-scale sequencing. *Nature*. 2010;467(7319):1061–1073. doi:10.1038/nature09534.

45. Prüfer K, Racimo F, Patterson N, et al. The complete genome sequence of a Neanderthal from the Altai Mountains. *Nature*. 2014;505(7481):43–49. doi:10

.1038/nature12886. Meyer M, Kircher M, Gansauge MT, et al. A high-coverage genome sequence from an archaic Denisovan individual. *Science*. 2012;338(6104):222–226. doi:10.1126/science.1224344.

46. Sankararaman S, Patterson N, Li H, Pääbo S, Reich D. The date of interbreeding between Neandertals and modern humans. *PLoS Genet*. 2012;8(10):e1002947. doi:10.1371/journal.pgen.1002947.

47. Reich D, Green RE, Kircher M, et al. Genetic history of an archaic hominin group from Denisova Cave in Siberia. *Nature*. 2010;468(7327):1053–1060. doi:10.1038/nature09710.

48. Juric I, Aeschbacher S, Coop G. The strength of selection against Neanderthal introgression. *PLoS Genet*. 2016;12(11):e1006340. doi:10.1371/journal.pgen.1006340. Sankararaman S, Mallick S, Patterson N, Reich D. The combined landscape of Denisovan and Neanderthal ancestry in present-day humans. *Curr Biol*. 2016;26(9):1241–1247. doi:10.1016/j.cub.2016.03.037.

49. Simonti CN, Vernot B, Bastarache L, et al. The phenotypic legacy of admixture between modern humans and Neandertals. *Science*. 2016;351(6274):737–741. doi:10.1126/science.aad2149.

50. Huerta-Sánchez E, Jin X, Asan, et al. Altitude adaptation in Tibetans caused by introgression of Denisovan-like DNA. *Nature*. 2014;512(7513):194–197. doi:10.1038/nature13408.

CHAPTER 6. SCIENCE VERSUS SCIENCE FICTION

1. Siegel V. I kid you not. *Dis Model Mech*. 2009;2(1–2):5–6. doi:10.1242/dmm.002352.

2. Hoffmann AA, Turelli M, Simmons GM. Unidirectional incompatibility between populations of *Drosophila simulans*. *Evolution*. 1986;40(4):692–701. doi:10.1111/j.1558-5646.1986.tb00531.x. Turelli M, Hoffmann AA. Cytoplasmic incompatibility in *Drosophila simulans*: Dynamics and parameter estimates from natural populations. *Genetics*. 1995;140(4):1319–1338.

3. Blagrove MS, Arias-Goeta C, Failloux AB, Sinkins SP. *Wolbachia* strain wMel induces cytoplasmic incompatibility and blocks dengue transmission in *Aedes albopictus*. *Proc Natl Acad Sci USA*. 2012;109(1):255–260. doi:10.1073/pnas.1112021108.

4. Aliota MT, Walker EC, Uribe Yepes A, Velez ID, Christensen BM, Osorio JE. The wMel strain of *Wolbachia* reduces transmission of chikungunya virus in *Aedes aegypti*. *PLoS Negl Trop Dis*. 2016;10(4):e0004677. doi:10.1371/journal.pntd.0004677.

5. Aliota MT, Peinado SA, Velez ID, Osorio JE. The wMel strain of *Wolbachia* reduces transmission of Zika virus by *Aedes aegypti*. *Sci Rep*. 2016;6:28792. doi:10.1038/srep28792.

6. Guzman MG, Halstead SB, Artsob H, et al. Dengue: A continuing global threat. *Nat Rev Microbiol.* 2010;8(Suppl. 12):S7–S16. doi:10.1038/nrmicro2460.

7. Murray JV, Jansen CC, De Barro P. Risk associated with the release of *Wolbachia*-infected *Aedes aegypti* mosquitoes into the environment in an effort to control dengue. *Front Public Health.* 2016;4:43. doi:10.3389/fpubh.2016.00043.

8. Schmidt TL, Barton NH, Rašić G, et al. Local introduction and heterogeneous spatial spread of dengue-suppressing *Wolbachia* through an urban population of *Aedes aegypti. PLoS Biol.* 2017;15(5):e2001894. doi:10.1371/journal.pbio.2001894.

9. Brock TD, Freeze H. *Thermus aquaticus* gen. n. and sp. n., a nonsporulating extreme thermophile. *J Bacteriol.* 1969;98(1):289–297.

10. Freeze H, Brock TD. Thermostable aldolase from *Thermus aquaticus. J Bacteriol.* 1970;101(2):541–550.

11. Saiki RK, Scharf S, Faloona F, et al. Enzymatic amplification of beta-globin genomic sequences and restriction site analysis for diagnosis of sickle cell anemia. *Science.* 1985;230(4732):1350–1354. doi:10.1126/science.2999980.

12. Miki Y, Swensen J, Shattuck-Eidens D, et al. A strong candidate for the breast and ovarian cancer susceptibility gene *BRCA1. Science.* 1994;266(5182):66–71. doi:10.1126/science.7545954.

13. Kornberg A. Basic research, the lifeline of medicine. Nobelprize.org. Nobel Media AB. http://www.nobelprize.org/nobel_prizes/medicine/laureates/1959/kornberg-article.html. Published July 23, 1997. Accessed January 8, 2018.

14. National Institute of General Medical Sciences. Curiosity creates cures: The value and impact of basic research. National Institute of General Medical Sciences website. https://www.nigms.nih.gov/education/pages/factsheet_curiosity createscures.aspx. Accessed January 8, 2018.

15. Salter AJ, Martin BR. The economic benefits of publicly funded basic research: A critical review. *Res Pol.* 2001;30(3):509–532. doi:10.1016/S0048-7333(00)00091-3.

16. Lyons T. Different countries, same science classes: Students' experiences of school science in their own words. *Int J Sci Educ.* 2006;28(6)591–613. doi:10.1080/09500690500339621.

17. Kramer M. Why we still love "Star Trek," final frontiers and all. Space.com. https://www.space.com/21162-star-trek-fan-love-endures.html. Published May 15, 2013. Accessed January 8, 2018.

18. Hawking S. Interview by Larry King. *Larry King Live,* CNN, December 25, 1999. Transcript available at https://www.nightscribe.com/Science_Technology/Science/hawking_LarryKing_Interview.htm.

19. Did science fiction influence you? Sigma Xi.org. https://www.sigmaxi.org/about/donate/did-science-fiction-influence-you. Accessed January 8, 2018.

20. Boutillette M, Coveney C, Kun S, Menides L. The influence of science fiction films on the development of biomedical instrumentation. *IEEE Xplore.* 2002; accession 6307371. doi:10.1109/NEBC.1999.755802.

21. Nesheim KC, Masner L, Johnson NF. The *Phanuromyia galeata* species group (Hymenoptera, Platygastridae, Telenominae): Shining a lantern into an unexplored corner of Neotropical diversity. *Zookeys*. 2017;(663):71–105. doi:10 .3897/zookeys.663.11554.

22. Erwin TL. Arboreal beetles of neotropical forests: *Agra* Fabricius, a taxonomic supplement for the *Platyscelis* group with new species and distribution records (Coleoptera: Carabidae, Lebiini, Agrina). *Coleop Bull*. 2000;54(1):90–119. doi:10.1649/0010–065X(2000)054[0090:ABONFA]2.0.CO;2.

23. Fraaije RHB, van Bakel BWM, Jagt JWM, Klompmaker AA, Artal P. A new hermit crab (Crustacea, Anomura, Paguroidea) from the mid-Cretaceous of Navarra, northern Spain. *Bol Soc Geol Mex*. 2009;61(2):211–214.

24. Paone Viegas DC, Pereira Benaim N, Silva Absãlo R. Description of four new species of *Ledella* Verrill and Bush, 1897 (Pelecypoda: Nuculanidae) off the coast of Brazil, using a morphometric approach. *Am Malacolog Bull*. 2014; 32(2):183–197. doi:10.4003/006.032.0201.

25. Bachtrog D. Accumulation of Spock and Worf, two novel non-LTR retrotransposons, on the neo-Y chromosome of *Drosophila miranda*. *Mol Biol Evol*. 2003;20(2):173–181. doi:10.1093/molbev/msg035.

26. Picard. Broad Institute website. https://broadinstitute.github.io/picard/. Accessed January 8, 2018.

27. Bixler A. Teaching evolution with the aid of science fiction. *Am Biol Teach*. 2007;69(6):337–340. doi:10.1662/0002–7685(2007)69[337:TEWTAO]2.0.CO;2.

APPENDIX: MINING GEMS AND COAL

1. Van de Wiele T, Van Praet JT, Marzorati M, Drennan MB, Elewaut D. How the microbiota shapes rheumatic diseases. *Nat Rev Rheumatol*. 2016;12(7):398–411. doi:10.1038/nrrheum.2016.85.

2. Kau AL, Ahern PP, Griffin NW, Goodman AL, Gordon JI. Human nutrition, the gut microbiome and the immune system. *Nature*. 2011;474(7351):327–336. doi:10.1038/nature10213.

3. Human Microbiome Project Consortium. Structure, function and diversity of the healthy human microbiome. *Nature*. 2012;486(7402):207–214. doi:10 .1038/nature11234.

4. D'Argenio V, Salvatore F. The role of the gut microbiome in the healthy adult status. *Clin Chim Acta*. 2015;451(A):97–102. doi:10.1016/j.cca.2015.01.003.

5. Foxman B. The epidemiology of urinary tract infection. *Nat Rev Urol*. 2010; 7(12):653–660. doi:10.1038/nrurol.2010.190.

6. Grube M, Cardinale M, De Castro JV, Müller H, Berg G. Species-specific structural and functional diversity of bacterial communities in lichen symbioses. *ISME J*. 2009;3(9):1105–1115. doi:10.1038/ismej.2009.63.

7. Hawksworth DL. The variety of fungal-algal symbioses, their evolutionary significance, and the nature of lichens. *Bot J Linn Soc.* 1988;96(1):3–20. doi:10.1111/j.1095–8339.1988.tb00623.x.

8. Margulis L. Symbiogenesis. A new principle of evolution rediscovery of Boris Mikhaylovich Kozo-Polyansky (1890–1957). *Paleontol J.* 2010;44(12):1525–1539. doi:10.1134/S0031030110120087.

9. Gill N, Findley S, Walling JG, et al. Molecular and chromosomal evidence for allopolyploidy in soybean. *Plant Physiol.* 2009;151(3):1167–1174. doi:10.1104/pp.109.137935. Cronn RC, Small RL, Wendel JF. Duplicated genes evolve independently after polyploid formation in cotton. *Proc Natl Acad Sci USA.* 1999;96(25):14406–14411. doi:10.1073/pnas.96.25.14406.

10. Hegarty MJ, Hiscock SJ. Genomic clues to the evolutionary success of polyploid plants. *Curr Biol.* 2008;18(10):R435–R444. doi:10.1016/j.cub.2008.03.043.

11. Chapman MA, Burke JM. Genetic divergence and hybrid speciation. *Evolution.* 2007;61(7):1773–1780. doi:10.1111/j.1558–5646.2007.00134.x.

12. Weiler N. Wasp virus turns ladybugs into zombie babysitters. *Science.* February 10, 2015. doi:10.1126/science.aaa7844.

13. Graham LE, Cook ME, Hanson DT, Pigg KB, Graham JM. Structural, physiological, and stable carbon isotopic evidence that the enigmatic Paleozoic fossil *Prototaxites* formed from rolled liverwort mats. *Am J Bot.* 2010;97(2):268–275. doi:10.3732/ajb.0900322.

14. Selosse MA. *Prototaxites*: A 400 myr old giant fossil, a saprophytic holobasidiomycete, or a lichen? *Mycol Res.* 2002;106(6):642–644. doi:10.1017/S0953756202226313. Graham LE, Cook ME, Hanson DT, Pigg KB, Graham JM. Structural, physiological, and stable carbon isotopic evidence that the enigmatic Paleozoic fossil *Prototaxites* formed from rolled liverwort mats. *Am J Bot.* 2010;97(2):268–275. doi:10.3732/ajb.0900322. Taylor TN, Taylor EL, Decombeix AL, et al. The enigmatic Devonian fossil *Prototaxites* is not a rolled-up liverwort mat: Comment on the paper by Graham et al. (*AJB* 97: 268–275). *Am J Bot.* 2010; 97(7):1074–1078. doi:10.3732/ajb.1000047.

15. Avery OT, MacLeod CM, McCarty M. Studies of the chemical nature of the substance inducing transformation of pneumococcal types. *J Exp Med.* 1944; 79(2):137–159. doi:10.1084/jem.79.2.137.

16. Dunning Hotopp JC, Clark ME, Oliveira DC, et al. Widespread lateral gene transfer from intracellular bacteria to multicellular eukaryotes. *Science.* 2007;317(5845):1753–1756. doi:10.1126/science.1142490.

17. Moran NA, Jarvik T. Lateral transfer of genes from fungi underlies carotenoid production in aphids. *Science.* 2010;328(5978):624–627. doi:10.1126/science.1187113.

18. Riley DR, Sieber KB, Robinson KM, et al. Bacteria-human somatic cell lateral gene transfer is enriched in cancer samples. *PLoS Comput Biol.* 2013;9(6): e1003107. doi:10.1371/journal.pcbi.1003107.

19. Boothby TC, Tenlen JR, Smith FW, et al. Evidence for extensive horizontal gene transfer from the draft genome of a tardigrade. *Proc Natl Acad Sci USA.* 2015;112(52):15976–15981. doi:10.1073/pnas.1510461112.

20. Koutsovoulos G, Kumar S, Laetsch DR, et al. No evidence for extensive horizontal gene transfer in the genome of the tardigrade *Hypsibius dujardini*. *Proc Natl Acad Sci USA*. 2016;113(18):5053–5058. doi:10.1073/pnas.1600338113.

21. Salzberg SL. Horizontal gene transfer is not a hallmark of the human genome. *Genome Biol*. 2017;18(1):85. doi:10.1186/s13059-017-1214-2. Martin WF. Too much eukaryote LGT. *Bioessays*. 2017;39(12):1700115 doi:10.1002/bies.2017 00115.

22. Smith FW, Bartels PJ, Goldstein B. A hypothesis for the composition of the tardigrade brain and its implications for panarthropod brain evolution. *Integr Comp Biol*. 2017;57(3):546–559. doi:10.1093/icb/icx081.

SUGGESTED ASSOCIATED SCIENTIFIC READING

CHAPTER 1

Gross, M. 2016. How life can arise from chemistry. *Current Biology* 26: R1247–R1249. doi:10.1016/j.cub.2016.12.001.

National Research Council. 2007. *The Limits of Organic Life in Planetary Systems*. National Academies Press. https://www.nap.edu/catalog/11919/the-limits-of-organic-life-in-planetary-systems.

Szostak, J. W. 2016. On the origin of life. *Medicina* (Buenos Aires) 76: 199–203.

CHAPTER 2

Coyne, J. A. 2009. *Why Evolution Is True*. Oxford University Press.

Gregory, T. R. 2008. Understanding evolutionary trees. *Evolution Education and Outreach* 1: 35. doi:10.1007/s12052-008-0035-x.

Harrison, T. 2010. Apes among the tangled branches of human origins. *Science* 327(5965): 532–534. doi:10.1126/science.1184703.

CHAPTER 3

Crick, F. 1970. Central dogma of molecular biology. *Nature* 227: 561–563. doi:10.1038/227561a0.

Doudna, J. A., and E. Charpentier. 2014. The new frontier of genome engineering with CRISPR-Cas9. *Science* 346(6213): 1258096. doi:10.1126/science.1258096.

Lane, M., R. L. Robker, and S. A. Robertson. 2014. Parenting from before conception. *Science* 345(6198): 756–760. doi:10.1126/science.1254400.

Miko, I. 2008. Gregor Mendel and the principles of inheritance. *Nature Education* 1(1): 134.

CHAPTER 4

Dawkins, R. 1996. *The Blind Watchmaker: Why the Evidence of Evolution Reveals a Universe without Design*. W. W. Norton.

Gregory, T. R. 2009. Understanding natural selection: Essential concepts and common misconceptions. *Evolution: Education and Outreach* 2: 128. doi:10.1007/s12052-009-0128-1.

Padian, K. 2013. Correcting some common misrepresentations of evolution in textbooks and the media. *Evolution: Education and Outreach* 6: 11. doi:10.1186/1936-6434-6-11.

Ségurel, L., and C. Bon. 2017. On the evolution of lactase persistence in humans. *Annual Review of Genomics and Human Genetics* 18: 392–319. doi:10.1146/annurev-genom-091416–035340.

CHAPTER 5

Bondar, C. 2016. *Wild Sex: The Science behind Mating in the Animal Kingdom*. Pegasus.

Johnson, N. A. 2008. Haldane's rule: The heterogametic sex. *Nature Education* 1(1): 58.

Johnson, N. A. 2008. Hybrid incompatibility and speciation. *Nature Education* 1(1): 20.

Otto, S. P. Sexual reproduction and the evolution of sex. *Nature Education* 1(1): 182.

Racimo, F., S. Sankaraman, R. Nielsen, and E. Huerta-Sanchez. 2015. Evidence for archaic adaptive introgression in humans. *Nature Reviews Genetics* 16(6): 359–371. doi:10.1038/nrg3936.

CHAPTER 6

Barad, J., and E. Robertson. 2001. *The Ethics of Star Trek*. Harper Perennial.

Brennan, P.L.R., D. Irschick, N. Johnson, and R. C. Albertson. 2014. Oddball science: Why studies of unusual evolutionary phenomena are crucial. *BioScience* 64: 178–179. doi:10.1093/biosci/bit039.

National Institute of General Medical Sciences. 2012. Curiosity creates cures: The value and impact of basic research. National Institute of General Medical Sciences website. https://www.nigms.nih.gov/Education/Pages/factsheet_CuriosityCreatesCures.aspx. Accessed January 8, 2018.

Siegel, E. 2017. *Treknology: The Science of Star Trek from Tricorder to Warp Drive*. Voyageur.

STAR TREK
EPISODE INDEX

SUBJECT INDEX

abalone, fertilization in, 123, 126
adenosine triphosphate (ATP), 23,
  24, 44
aerobic respiration, 9, 23, 24, 45; by
  tardigrades, 155
aging: in *Star Trek*, 61, 63, 75, 92;
  telomeres and, 92
Agra dax, 140
algae, of lichens, 145
allopolyploidy, 147
Amazon molly, 112
amino acids: chirality of, 17; extraterrestrially formed, 16, 43; genetic
  code and, 32–33; origin of life and,
  6, 16, 43n; proteins and, 61, 62,
  63, 67
ammonia, 13
anaerobic respiration, 23–24, 27
Andorians, 1
android. *See* Data, Lieutenant
  Commander
Annutidiogenes worfi, 140
antibiotic resistance, 3, 90–91
antibiotics: killing normal body flora,
  145; wasp hybridization and, 124
antiprotons, 77
aphids, gene transfer into, 152–53
archaea: anaerobic, 23–24; heat-
  tolerant, 20
Archer, Jonathan, 131

Archer, Karyn, 131
asexual reproduction, 106, 107, 108,
  109–10; evolved from sexual species, 109, 112; in *Star Trek*, 110; of
  Tetrahymena's seven mating types,
  115. *See also* clones
Asimov, Isaac, 14–15, 27
asteroid impact, 44, 45, 50, 97
astronauts: alien, 46–49; as fans of
  Star Trek, 139
ATP (adenosine triphosphate), 23,
  24, 44
Axanar, 26

bacteria: alive in ancient ice cores,
  21; anaerobic respiration in, 23, 27;
  antibiotics and, 3, 90–91, 124, 145;
  with carbon-silicon bonds, 17; chemosynthesis in, 25; endosymbiotic
  theory and, 45–46; as example of
  life, 3; of human microbiome, 145;
  intracellular, in wasp species, 124;
  panspermia and, 43; PCR using
  enzyme from, 136–37; photosynthesis in, 24, 25; powered by radio-
  activity, 25; in *Star Trek* episode,
  27; transferring or taking in DNA,
  108, 152, 153; *Wolbachia* used in
  disease control, 135–36, 152. *See
  also* microbes